# Advanced Digital Troubleshooting

# Advanced Digital Troubleshooting

**Alvis J. Evans**

HOWARD W. SAMS & COMPANY
*A Division of Macmillan, Inc.*
*4300 West 62nd Street*
*Indianapolis, Indiana 46268 USA*

FIRST EDITION
FIRST PRINTING--1988



International Standard Book Number: 0-672-22571-9
Library of Congress Catalog Card Number: 88-60310

Acquisitions Editor: *Greg Michael*
Developmental Editor: *Jennifer Ackley*
Manuscript Editor: *Technical Editing Service, Tom Whipple*
Illustrators: *T. R. Emrick and David Cripe*
Cover Art: *Laura Gibbons*
Indexer: *Don Herrington*
Composition: *BMEP, Inc., Zionsville, IN*

*Printed in the United States of America*

**Trademark Acknowledgments**

*DEDICATED*

*to my father, Raymond Evans*

# Contents

# Preface

*Advanced Digital Troubleshooting* emphasizes the practical techniques of troubleshooting digital equipment, particularly the wide variety of microprocessor-based systems. Electronics technicians had to change tactics as solid-state technology replaced vacuum tubes. Likewise, the digital technician must use methods different from those used by a technician working on linear circuits. Though this is not a book on digital and microprocessor theory, sufficient introductory discussion is provided for an understanding necessary to master the troubleshooting techniques presented. Since digital and microprocessor circuits are found from the factory to the kitchen and from the automobile to the child's toy, the ability to analyze circuits systematically, locate faults quickly, and repair them effectively is the most universal skill demanded by employers of electronics technicians.

The book begins with a review of the electrical characteristics of digital systems and progresses through hardware and software concepts of microprocessor technology, taking a modern approach to troubleshooting.

Though it is popular to emphasize replacement of printed circuit boards and troubleshooting to the board level only, our world is presently stirring with a new awareness of recycling our resources of all kinds. The economic life of any product or machine begins when it is first placed in service and ends when the cost of repair and maintenance becomes greater than the cost of replacement. This certainly will become more of a factor as to whether to troubleshoot to the board or to the component level. The basis for making this decision will, most probably, depend on the effectiveness of the troubleshooter in quickly finding the faulty component and returning the equipment to service as fast as possible. This concept is emphasized throughout the book.

My thanks go to the great people that I work with at Tarrant County Junior College: the instructional team and the support staff of the electronics department.

Modern troubleshooting would be impossible without modern test equipment. I have had tremendous support on state-of-the-art equipment from Gary Woltal and Bruce Sundquist at Tektronics, Inc., and from Craig Hendrix at Hewlett-Packard.

A special thanks to my son Chad for preparing the many drawings and taking the photographs.

And finally, I thank Darlene, my soul mate, who not only makes this book possible, but more importantly, makes it and *all else* worthwhile.

ALVIS J. EVANS

# 1
# Electrical Characteristics of Digital Systems

In this chapter, basic digital logic circuits are briefly reviewed, and then digital electrical parameters are discussed in detail. Common terms are reviewed and then related to manufacturers' data sheets. Both static and dynamic parameters of digital devices are explained. Basic digital pulse-measuring techniques are introduced along with the terms commonly used in describing pulses found in digital circuits.

## Basic Digital Logic Circuits—a Review

In order to perform troubleshooting of digital circuits and systems, the technician must be able to read and understand logic symbols. Although we assume that the technician is already acquainted with basic digital logic, this review will familiarize all readers with the common symbols and terms used in industrial digital electronics. Most of the discrete logic circuits in digital equipment consist of two general types of logic circuits: combinational and sequential.

### Combinational Logic Circuits

In a combinational logic circuit, the output depends on the input combination present at the time that the output is examined. The decision-making capability of combinational circuits is the primary reason for their wide application.

All digital circuit functions are accomplished through one or more of the logic operations AND, OR, and NOT. The devices that perform these operations are called *gates*. The symbols for these three basic gates with their truth tables are given in Figure 1-1. These three logic operators can be connected so that all other possible switching functions can be realized.

Figure 1-2 shows other logic functions. with their truth tables, that may be formed from the basic three. These logic circuits perform logic operations according to the rules of a mathematical system known as *Boolean algebra*. The Boolean algebra expression for each logic function is shown in the figure as the gate output. Since Boolean algebra expressions are often complex, the logic function can be expressed much simpler as a truth table—that is, a systema-

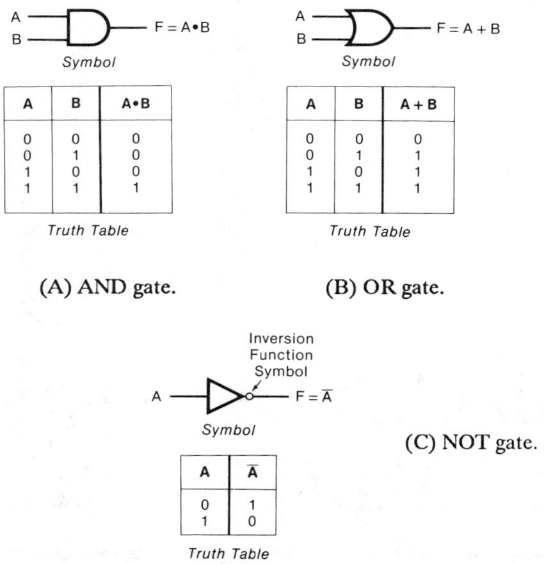


| A | B | A•B |
|---|---|---|
| 0 | 0 | 0 |
| 0 | 1 | 0 |
| 1 | 0 | 0 |
| 1 | 1 | 1 |

| A | B | A+B |
|---|---|---|
| 0 | 0 | 0 |
| 0 | 1 | 1 |
| 1 | 0 | 1 |
| 1 | 1 | 1 |

| A | $\overline{A}$ |
|---|---|
| 0 | 1 |
| 1 | 0 |

**Figure 1-1.** The logic symbol, Boolean algebra expression, and truth table for the three basic logic functions.

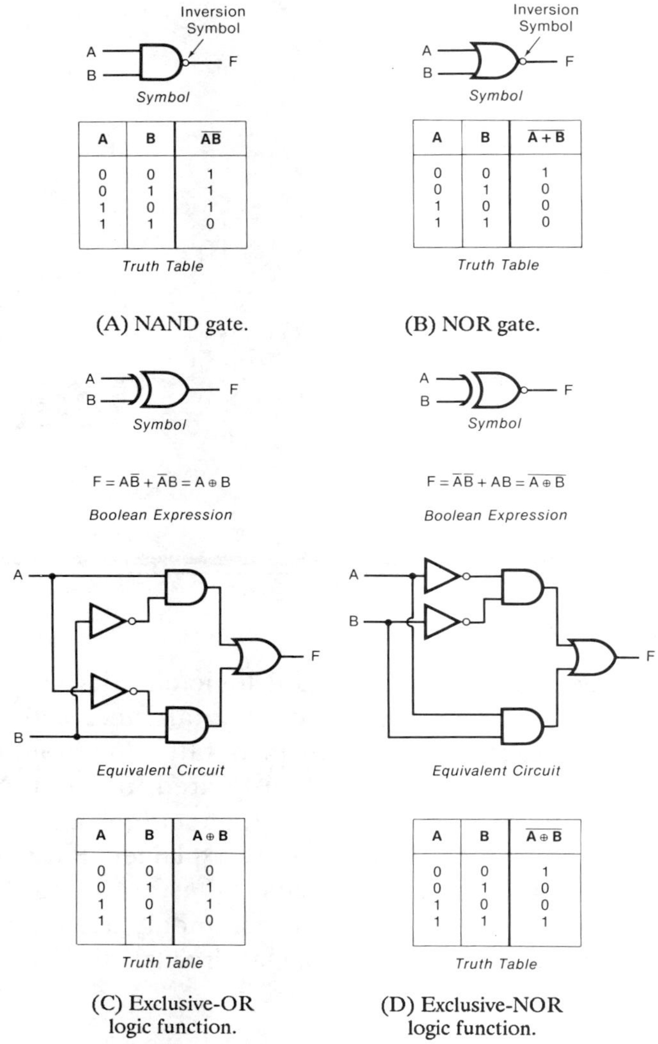


| A | B | $\overline{AB}$ |
|---|---|---|
| 0 | 0 | 1 |
| 0 | 1 | 1 |
| 1 | 0 | 1 |
| 1 | 1 | 0 |

| A | B | $\overline{A+B}$ |
|---|---|---|
| 0 | 0 | 1 |
| 0 | 1 | 0 |
| 1 | 0 | 0 |
| 1 | 1 | 0 |

| A | B | $A \oplus B$ |
|---|---|---|
| 0 | 0 | 0 |
| 0 | 1 | 1 |
| 1 | 0 | 1 |
| 1 | 1 | 0 |

| A | B | $\overline{A \oplus B}$ |
|---|---|---|
| 0 | 0 | 1 |
| 0 | 1 | 0 |
| 1 | 0 | 0 |
| 1 | 1 | 1 |

**Figure 1-2.** Other logic functions.

tic listing of all possible combinations of the states of the input variables, where each variable must be 0 or 1—with the corresponding outputs assigned to each combination.

Two of the most important laws of Boolean algebra are DeMorgan's laws. They allow for easy transfer of the Boolean expression and circuit logic back and forth between OR (NOR) and AND (NAND) forms. Figure 1-3 shows the Boolean identity expressions, circuit equivalency, and the truth tables of these laws. The first law changes the basic OR situation to an AND situation, and the second law changes AND to OR. A combinational logic circuit is a network of logic gates which are interconnected to perform a desired function.

Now, as an example of what we have discussed, we will convert a circuit that performs a certain logic function using the three basic gates into a circuit that uses only NAND gates and performs the same logic function. The NAND gate has traditionally been the universal gate in digital electronic circuits. The original logic circuit and its output function are shown in Figure 1-4A. The equivalent NAND logic circuit is shown in Figure 1-4C, with Figure 1-4B showing the intermediate step. Notice that the inverter of Figure 1-4A has been replaced by a NAND gate with its input terminals tied together. A look at a truth table will verify that this is equivalent to an inverter. Two inverters are added between gates 2 and 4 and between gates 3 and 4. Double inversion of the signal out of gates 2 and 3 and into gate 4 has no effect on the signals. However, by DeMorgan's second law, gates 2, 3, and 4 may now be replaced by NAND gates. Figure 1-4A contains three different kinds of gates requiring three separate inefficiently used chips, whereas Figure 1-4C requires only one chip that is fully utilized.

| A | B | $\overline{A+B}$ |
|---|---|---|
| 0 | 0 | 1 |
| 0 | 1 | 0 |
| 1 | 0 | 0 |
| 1 | 1 | 0 |

*Truth Table*

$\overline{A+B} \equiv \bar{A} \cdot \bar{B}$

*Boolean Expression*

| A | B | $\bar{A} \cdot \bar{B}$ |
|---|---|---|
| 0 | 0 | 1 |
| 0 | 1 | 0 |
| 1 | 0 | 0 |
| 1 | 1 | 0 |

*Truth Table*

NOTE: ≡ Means "Is Identical to".

(A) DeMorgan's first law.

| A | B | $\overline{A \cdot B}$ |
|---|---|---|
| 0 | 0 | 1 |
| 0 | 1 | 1 |
| 1 | 0 | 1 |
| 1 | 1 | 0 |

*Truth Table*

$\overline{A \cdot B} \equiv \bar{A} + \bar{B}$

*Boolean Expression*

| A | B | $\bar{A} + \bar{B}$ |
|---|---|---|
| 0 | 0 | 1 |
| 0 | 1 | 1 |
| 1 | 0 | 1 |
| 1 | 1 | 0 |

*Truth Table*

(B) DeMorgan's second law.

**Figure 1-3.** Equivalent forms of DeMorgan's laws.

Combinational logic is that logic which, neglecting transit time, responds immediately to a combination of inputs. We will now look at the other extremely valuable type of logic circuit.

## Sequential Logic Circuits

The basic building block of combinational logic is the logic gate. The basic building block of the sequential logic circuit is the *flip-flop* (FF) circuit. The value of the sequential logic circuit is its memory characteristic. Flip-flops (also called latches and multivibrators) can be made from basic gates, such as NAND gates. However, they are usually bought in integrated circuit form. In digital applications, flip-flops are usually interconnected to form sequential circuits that are used for storing data (memory), counting, timing, or sequencing data.

Sequential circuits are either clocked (referred to as synchronous) or unclocked (asynchronous). In synchronous logic, there is a master oscillator called a *clock* that provides regular timing pulses. Events are permitted to occur only during one of these timing pulses. The clock pulse is sometimes referred to as a *strobe*. In an asynchronous FF, the outputs are permitted to change as the inputs change without having to wait for a clock pulse to strobe in new data.

The most basic FF is called the SR flip-flop. Figure 1-5 shows the block logic symbol, the circuit connections using two NAND gates, and its truth table. The SR FF in this example has two inputs labeled SET (S) and RESET (R), shown with small circles, which means they are active low inputs. Unlike the logic gates, the FF has two complementary outputs, usually labeled Q and $\overline{Q}$ (Q NOT).

In the truth table the first line is called the prohibited state, because it would drive both outputs high (i.e., their values would be 1s). An SR FF is typically used to temporarily hold data. Hence, it is often called a latch. The basic SR FF is an asynchronous device because it has no provisions for a clock pulse input. The addition

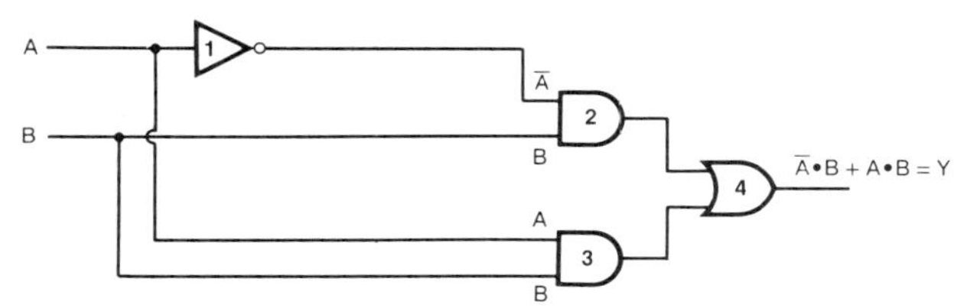


(A) An AND/OR logic circuit.

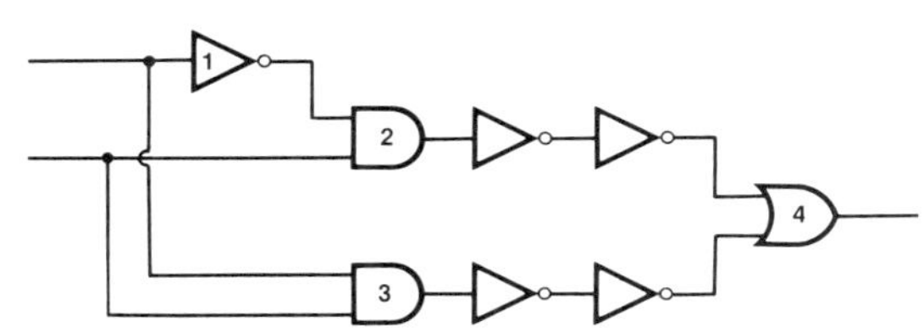


(B) Intermediate step between AND/OR and NAND logic circuit.

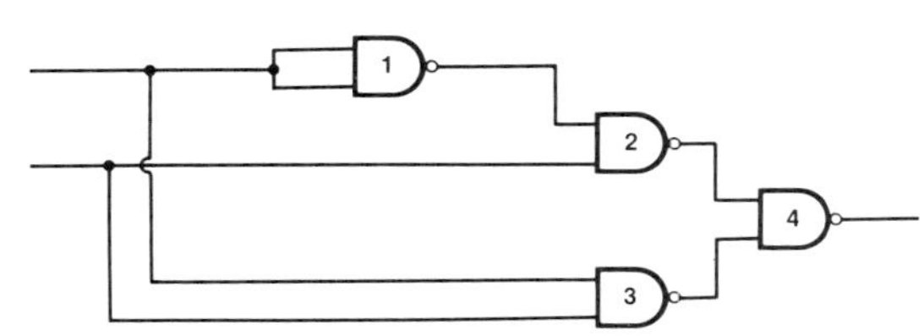


(C) A NAND logic circuit.

**Figure 1-4.** Converting an AND/OR logic circuit to a NAND logic circuit.

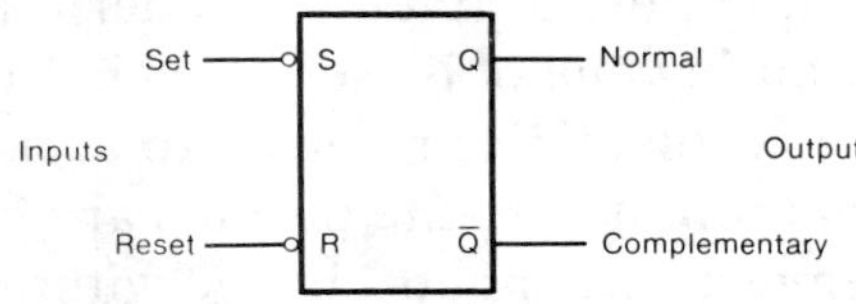


(A) Logical symbol for SR FF.

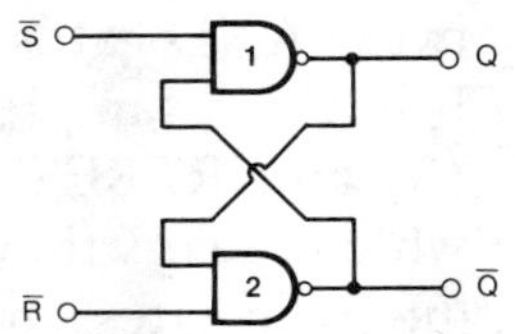


(B) Wired SR FF using NAND gate.

| S̄ | R̄ | Q |
|---|---|---|
| 0 | 0 | Forbidden |
| 0 | 1 | 1 |
| 1 | 0 | 0 |
| 1 | 1 | No Change |

(C) Truth table.

**Figure 1-5.** SR flip-flop.

of this clock or strobe input adds the valuable synchronous feature to the SR latch.

Figure 1-6 shows the basic clocked SR FF. Notice that two NAND gates (3 and 4) have been added to the basic SR latch to form this clocked SR FF. Because of the inverting effect of these gates, S and R are active high inputs. The SR latches we have seen are called *level triggered* because the information at the data inputs will be transferred to the outputs anytime during the clock pulse interval. The data must not be allowed to change during the strobe period or improper operation will result. More commonly, edge triggering is used for most FFs. This type of strobe is such that data are allowed to be transferred from the input of the FF or latch to the output only during the transition of the strobe pulse from low to high, or vice versa. We will discuss edge triggering further when we consider the two most useful and versatile FFs.

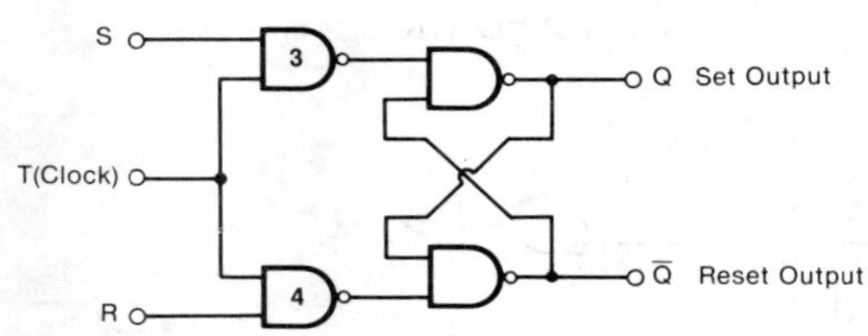


(A) Circuit.

| Input S | Input R | Input T | Command | Output Q | Output Q̄ |
|---|---|---|---|---|---|
| L | L | L | | NC | NC |
| L | L | H | Remember | NC | NC |
| L | H | L | | NC | NC |
| L | H | H | Reset | L | H |
| H | L | L | | NC | NC |
| H | L | H | Set | H | L |
| H | H | L | | NC | NC |
| H | H | H | Invalid | H | H |

Note: NC = No Change
L = Low
H = High

(B) Truth table.

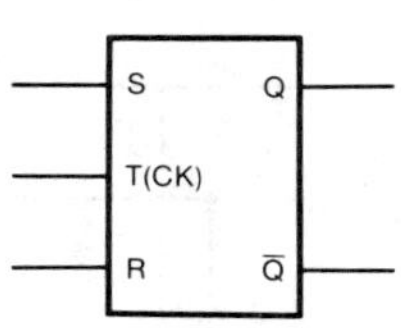


(C) Symbol.

**Figure 1-6.** Basic clock SR flip-flop.

## D and JK Flip-Flops

A modification to the clocked SR FF will produce the D FF as shown in Figure 1-7. Note

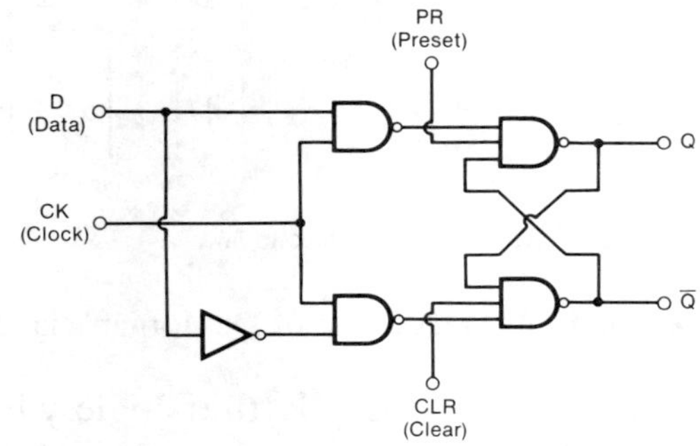


(A) Circuit.

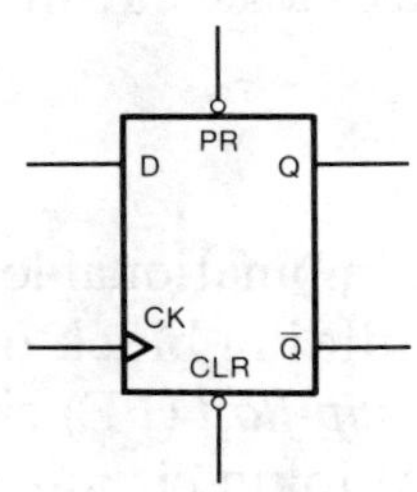


(B) Symbol.

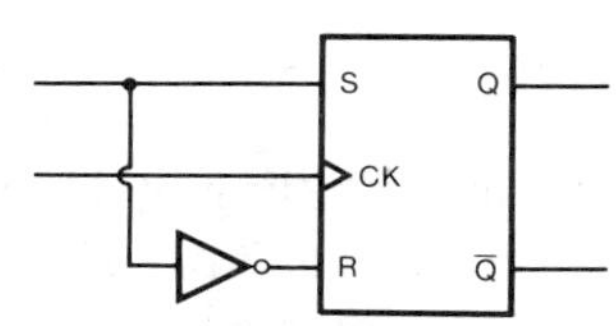


(C) SR FF wired to form D FF.

| Mode of Operation | Inputs: Asynchronous PR | Inputs: Asynchronous CLR | Inputs: Synchronous CK | Inputs: Synchronous D | Outputs Q | Outputs Q̄ |
|---|---|---|---|---|---|---|
| Asynchronous Set | 0 | 1 | X | X | 1 | 0 |
| Asynchronous Reset | 1 | 0 | X | X | 0 | 1 |
| Prohibited | 0 | 0 | X | X | 1 | 1 |
| Set | 1 | 1 | ↑ | 1 | 1 | 0 |
| Reset | 1 | 1 | ↑ | 0 | 0 | 1 |

0 = low, 1 = high, X = irrelevant,
↑ = low-to-high transition of the clock pulse

(D) Truth table.

**Figure 1-7.** D flip-flop.

the addition of the inverter. Also note the clock (CK) input. The small wedge inside the logic symbol for the D FF means that this is an edge-triggered device as opposed to a level-triggered device. The truth table for the D FF (7474) shows both synchronous and asynchronous operation. The first three lines are for asynchronous operation using the preset (PR) and clear (CLR) inputs. With the PR and CLR inputs high (disabled), the synchronous inputs (D and CK) will operate. Only the bottom two lines of the truth table are used for synchronous operation. Because D FFs are widely used for data storage, they are sometimes referred to as data FFs.

The JK FF has all the synchronous qualities of the clocked SR plus a guaranteed valid output for all possible inputs. The JK FF logic symbol and mode (truth) table are shown in Figure 1-8. The modes of operation are given in the left column. Notice that a useful toggle FF (T FF) results from tying both J and K inputs high. When repeated clock pulses are applied to the CK input, the output will go to the opposite state with the arrival of each pulse; that is, it simply toggles.

A *counter* is a special-purpose sequential digital circuit composed of FFs whose function is to count any number of events required by the system specifications. The counter can be designed to provide an output pulse after any predetermined number of input pulses (that is, after it has reached the Nth count). So, the frequency of the output pulses is equal to the frequency of the input pulses divided by N.

To clarify the discussion of sequential circuits, we often construct timing diagrams to

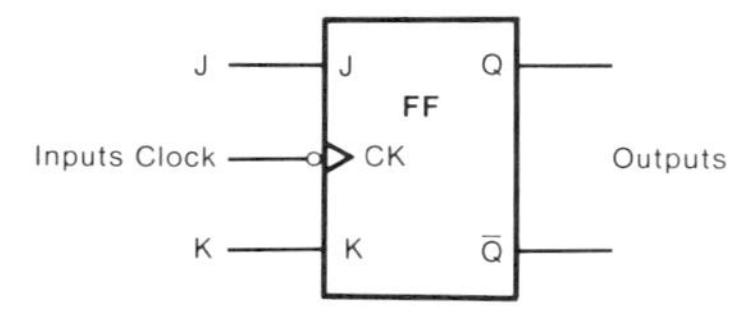


(A) Logic symbol.

| Mode of Operation | Inputs | | | Outputs | |
|---|---|---|---|---|---|
| | CK | J | K | Q | $\overline{Q}$ |
| Hold | ⎍ | 0 | 0 | No Change | |
| Reset | ⎍ | 0 | 1 | 0 | 1 |
| Set | ⎍ | 1 | 0 | 1 | 0 |
| Toggle | ⎍ | 1 | 1 | Opposite State | |

(B) Mode truth table for pulse-triggered JK FF.

**Figure 1-8.** JK flip-flop.

show the time relationship of voltages at various points in the circuit. Figure 1-9 illustrates a ripple up counter using D FFs. Notice that only the least significant bit (LSB) FF is strobed from the clock, which then strobes the next FF, and so on, through each FF. Thus, it is considered asynchronous or a ripple counter. Due to propagation time through each FF, the outputs switch at different times, resulting in a rather slow operation.

A parallel or synchronous counter, as shown in Figure 1-10, has the clock signal applied to all the FFs at once. The combination logic gates control the state changes of the various FFs. This counter does not have the propagation delay of the ripple counter and is more suitable for faster operation.

In addition to logic functions and counting, digital circuits also perform other arithmetic operations. All of these functions are performed in a computer's central processing unit (CPU),

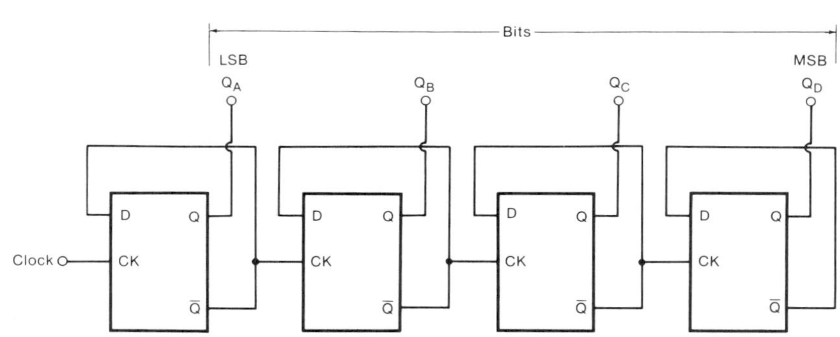


**Figure 1-9.** Ripple up counter using D flip-flops.

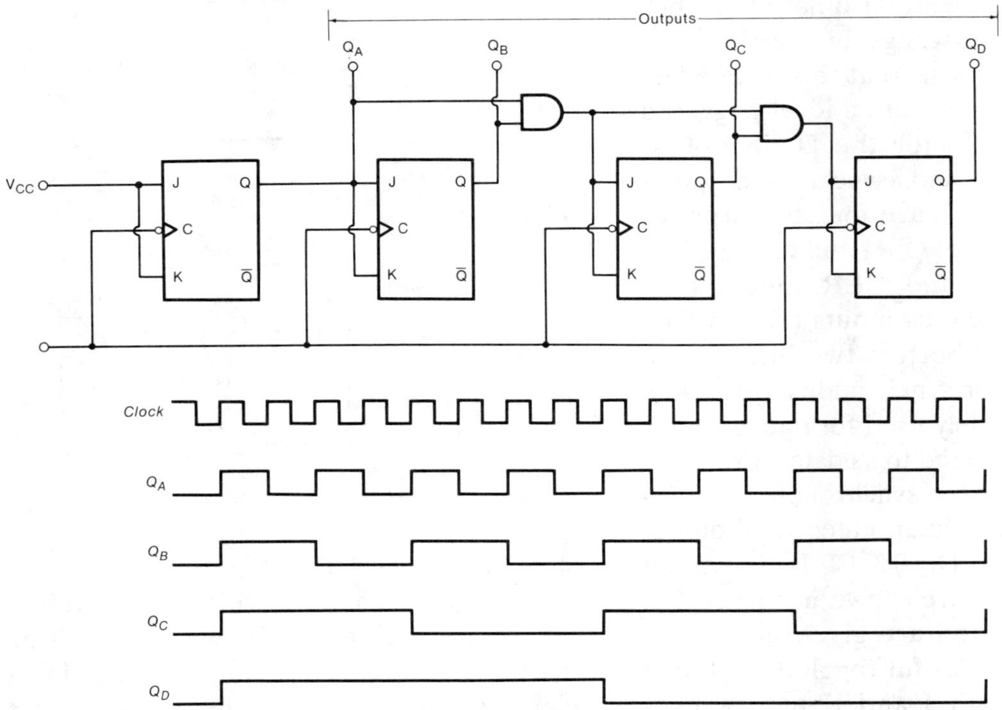


(A) Circuit diagram.

(B) Input and output waveforms.

**Figure 1-10.** Parallel or synchronous counter.

which contains the arithmetic logic unit (ALU). The CPU is discussed in Chapter 4.

## Digital Electrical Parameters

In a first course in digital logic, the circuits and signals are often treated as if they were ideal or perfect. Both are far from ideal, of course, but we will look specifically at two broad parameters: the signal voltage levels, and the circuit or device speed.

First, let's look at the reason that the signal levels are not ideal. It is not necessary to know how integrated circuits (ICs) are internally constructed in order to use them. It is extremely helpful to know the operating characteristic of each family to effectively troubleshoot the digital circuits. Just how important is it for the digital troubleshooter to know what is inside the IC? Are certain types more prone to specific failures? And, perhaps what is more important, can we, as technicians, accidentally damage a circuit by doing something that is normally safe for a different type of circuit?

There are two broad categories of chip fabrication processes: bipolar and metal-oxide semiconductor (MOS). Bipolar families depend on bipolar transistors to perform their digital operations, whereas MOS families depend on field-effect transistors (FETs). Generally, MOS ICs require fewer manufacturing steps, cover a smaller area on the substrate, and consume less power than do their bipolar equivalents. MOS devices usually operate at slower speeds, are more expensive, and require special handling and storage techniques. The families are listed in Table 1-1 under each category with some of their more important parameters.

Logic circuits are not uniform devices delivering precise voltage levels for logic 1 or 0. The power supply voltages $V_{CC}$ and $V_{DD}$ are usually presented as the ideal 1 level, whereas a 0-V ground is the ideal 0 level. Unfortunately, logic circuits are imperfect. We will see shortly that 1-level outputs are significantly lower than the power supply voltage, and 0-level outputs are

**Table 1-1. A Comparison Among Logic Families**

| Family | Speed (nSec) | Power Dissipated (mW) | Required Voltages (V) | High Logical Level (V) | Low Logical Level |
|---|---|---|---|---|---|
| RTL | 15 | 24 | +3.6 | +3.6 | 1.4 |
| DTL | 30 | 10 | +5 | +5 | 0.98 |
| TTL | 10 | 10 | +5 | +5 | 0.58 |
| ECL | 2 | 40 | - 5.2 | - 0.9 | -1.75 |
| CMOS | 25 | 0.00001 | +3 to +15 | $V_{cc}$ | 1.17 |
| Low-Speed TTL | 33 | 1 | +5 | +3 | 0.93 |
| Schottky TTL | 3 | 19 | +5 | +3 | 1.39 |
| Low-Power Schottky TTL | 9 | 2 | +5 | +3 | 1.39 |

significantly higher than ground level. In fact, the minimum 1 level and maximum 0 level will vary with the load connected to the output.

Two types of loads may be applied to a logic circuit: the sink load and the source load. A sink load is best represented as a sink current that an input will draw when driven by a specified 1 output level of the gate driving it. It therefore tends to degrade the 1 level from the output of the driving gate because of the voltage drop across the upper transistor and diode. A source load, on the other hand, does not degrade the 1 level but raises the 0 level of the driving gate. This voltage increase is caused by the voltage drop across the diode in the gate input. It resembles a higher resistance source. The concepts of sink and source are illustrated in Figure 1-11.

The current direction is assumed to be electron flow, and the gates represented are transistor-transistor logic (TTL). The sink load or source load is considered the "unit" load, that is, a load representative of the typical logic unit input. It is a general working specification that can be computed for convenience from either the manufacturer's specification sheet or the schematic drawing of the circuit. For a TTL 7400 series, one unit load (UL) is a source of 0.40 mA or 400 $\mu$A at the high level and a sink of 1.6 mA at the high level. The number of ULs the output of a gate can safely drive is its *fan-out*. The number of ULs the input will present to the output of another gate is its *fan-in*.

The voltage ranges of logical 1 and logical 0 input values are illustrated in Figure 1-12. The troubleshooter must be familiar with the electrical parameters of a device because trouble in a digital circuit is often caused by a device that does not meet its specified electrical parameters.

Much information is given in the number of the device. The technician should be aware of the method to the madness of the numbering system. Many times, the number is all that is given in service literature, somewhat like a nickname that presumes mutual knowledge on the part of the communicators. For example, one might call for a 7400 IC, when the actual chip had the number MC5400H-02 or F74H00P-01. What's the difference? Let's look at the chip numbering system by breaking it into four parts: prefix, unique circuit description (series), package, and dash number (instructions).

The prefix identifies the manufacturer and possibly some special manufacturer's information. Some identifying prefixes are MC

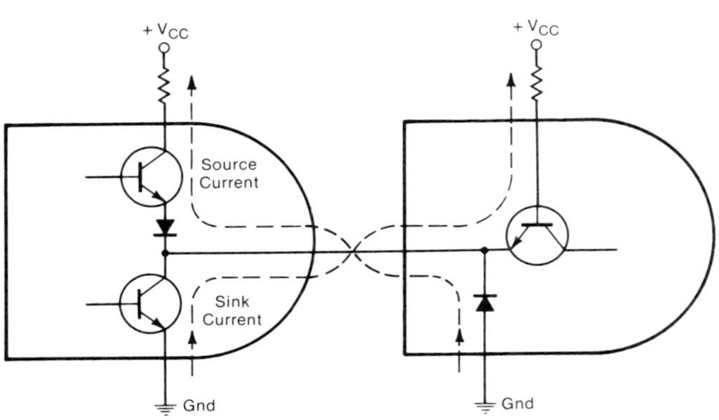


**Figure 1-11.** The sink and source concepts.

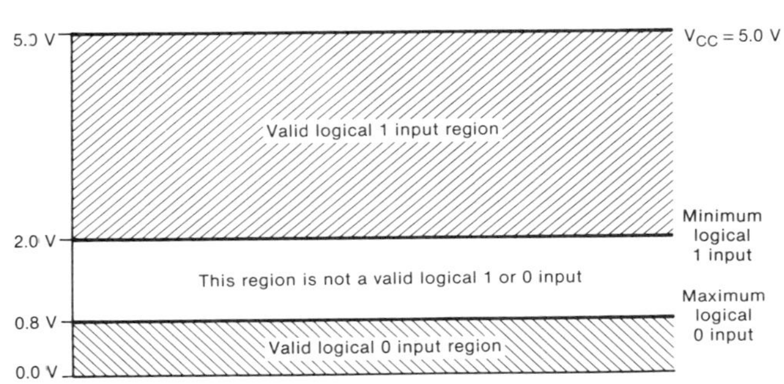


**Figure 1-12.** Boundaries for valid input voltages equal to logical 1 and logical 0.

**Table 1-2. Data Sheet of Terms and**

### recommended operating conditions

| | 54 FAMILY / 74 FAMILY | SERIES 54 / SERIES 74 '00, '04, '10, '20, '30 MIN | NOM | MAX | SERIES 54H / SERIES 74H 'H00, 'H04, 'H10, 'H20, 'H30 MIN | NOM | MAX | SERIES 54L / SERIES 74L 'L00, 'L04, 'L10, 'L20, 'L30 MIN | NOM | MAX | SERIES 54LS / SERIES 74LS 'LS00, 'LS04, 'LS10, 'LS20, 'LS30 MIN | NOM | MAX | SERIES 54S / SERIES 74S 'S00, 'S04, 'S10, 'S20, 'S30, 'S133 MIN | NOM | MAX | UNIT |
|---|---|---|---|---|---|---|---|---|---|---|---|---|---|---|---|---|---|
| Supply voltage, $V_{CC}$ | 54 Family | 4.5 | 5 | 5.5 | 4.5 | 5 | 5.5 | 4.5 | 5 | 5.5 | 4.5 | 5 | 5.5 | 4.5 | 5 | 5.5 | V |
| | 74 Family | 4.75 | 5 | 5.25 | 4.75 | 5 | 5.25 | 4.75 | 5 | 5.25 | 4.75 | 5 | 5.25 | 4.75 | 5 | 5.25 | |
| High-level output current, $I_{OH}$ | 54 Family | | | −400 | | | −500 | | | −100 | | | −400 | | | −1000 | μA |
| | 74 Family | | | −400 | | | −500 | | | −200 | | | −400 | | | −1000 | |
| Low-level output current, $I_{OL}$ | 54 Family | | | 16 | | | 20 | | | 2 | | | 4 | | | 20 | mA |
| | 74 Family | | | 16 | | | 20 | | | 3.6 | | | 8 | | | 20 | |
| Operating free-air temperature, $T_A$ | 54 Family | −55 | | 125 | −55 | | 125 | −55 | | 125 | −55 | | 125 | −55 | | 125 | °C |
| | 74 Family | 0 | | 70 | 0 | | 70 | 0 | | 70 | 0 | | 70 | 0 | | 70 | |

### electrical characteristics over recommended operating free-air temperature range (unless otherwise noted)

| PARAMETER | TEST FIGURE | TEST CONDITIONS† | | SERIES 54 / SERIES 74 '00, '04, '10, '20, '30 MIN | TYP‡ | MAX | SERIES 54H / SERIES 74H 'H00, 'H04, 'H10, 'H20, 'H30 MIN | TYP‡ | MAX | SERIES 54L / SERIES 74L 'L00, 'L04, 'L10, 'L20, 'L30 MIN | TYP‡ | MAX | SERIES 54LS / SERIES 74LS 'LS00, 'LS04, 'LS10, 'LS20, 'LS30 MIN | TYP‡ | MAX | SERIES 54S / SERIES 74S 'S00, 'S04, 'S10, 'S20, 'S30, 'S133 MIN | TYP‡ | MAX | UNIT |
|---|---|---|---|---|---|---|---|---|---|---|---|---|---|---|---|---|---|---|---|
| $V_{IH}$ High-level input voltage | 1, 2 | | | 2 | | | 2 | | | 2 | | | 2 | | | 2 | | | V |
| $V_{IL}$ Low-level input voltage | 1, 2 | | 54 Family | | | 0.8 | | | 0.8 | | | 0.7 | | | 0.7 | | | 0.8 | V |
| | | | 74 Family | | | 0.8 | | | 0.8 | | | 0.7 | | | 0.8 | | | 0.8 | |
| $V_{IK}$ Input clamp voltage | 3 | $V_{CC}$ = MIN, $I_I$ = § | | | | −1.5 | | | −1.5 | | | | | | −1.5 | | | −1.2 | V |
| $V_{OH}$ High-level output voltage | 1 | $V_{CC}$ = MIN, $V_{IL}$ = $V_{IL}$ max, $I_{OH}$ = MAX | 54 Family | 2.4 | 3.4 | | 2.4 | 3.5 | | 2.4 | 3.3 | | 2.5 | 3.4 | | 2.5 | 3.4 | | V |
| | | | 74 Family | 2.4 | 3.4 | | 2.4 | 3.5 | | 2.4 | 3.2 | | 2.7 | 3.4 | | 2.7 | 3.4 | | |
| $V_{OL}$ Low-level output voltage | 2 | $V_{CC}$ = MIN, $V_{IH}$ = 2 V, $I_{OL}$ = MAX | 54 Family | | 0.2 | 0.4 | | 0.2 | 0.4 | | 0.15 | 0.3 | | 0.25 | 0.4 | | | 0.5 | V |
| | | | 74 Family | | 0.2 | 0.4 | | 0.2 | 0.4 | | 0.2 | 0.4 | | 0.25 | 0.5 | | | 0.5 | |
| | | $I_{OL}$ = 4 mA | Series 74LS | | | | | | | | | | | | 0.4 | | | | |
| $I_I$ Input current at maximum input voltage | 4 | $V_{CC}$ = MAX | $V_I$ = 5.5 V | | | 1 | | | 1 | | | 0.1 | | | | | | 1 | mA |
| | | | $V_I$ = 7 V | | | | | | | | | | | | 0.1 | | | | |
| $I_{IH}$ High-level input current | 4 | $V_{CC}$ = MAX | $V_{IH}$ = 2.4 V | | | 40 | | | 50 | | | 10 | | | | | | | μA |
| | | | $V_{IH}$ = 2.7 V | | | | | | | | | | | | 20 | | | 50 | |
| $I_{IL}$ Low-level input current | 5 | $V_{CC}$ = MAX | $V_{IL}$ = 0.3 V | | | | | | | | | −0.18 | | | | | | | mA |
| | | | $V_{IL}$ = 0.4 V | | | −1.6 | | | −2 | | | | | | −0.4 | | | | |
| | | | $V_{IL}$ = 0.5 V | | | | | | | | | | | | | | | −2 | |
| $I_{OS}$ Short-circuit output current♦ | 6 | $V_{CC}$ = MAX | 54 Family | −20 | | −55 | −40 | | −100 | −3 | | −15 | −20 | | −100 | −40 | | −100 | mA |
| | | | 74 Family | −18 | | −55 | −40 | | −100 | −3 | | −15 | −20 | | −100 | −40 | | −100 | |
| $I_{CC}$ Supply current | 7 | $V_{CC}$ = MAX | | See table on next page | | | | | | | | | | | | | | | mA |

†For conditions shown as MIN or MAX, use the appropriate value specified under recommended operating conditions.
‡All typical values are at $V_{CC}$ = 5 V, $T_A$ = 25°C.
§$I_I$ = −12 mA for SN54'/SN74', −8 mA for SN54H'/SN74H', and −18 mA for SN54LS'/SN74LS' and SN54S'/SN74S'.
♦Not more than one output should be shorted at a time, and for SN54H'/SN74H', SN54LS'/SN74LS', and SN54S'/SN74S', duration of short-circuit should not exceed 1 second.

**Electrical Parameters** *(Courtesy Texas Instruments Inc.)*

## supply current¶

| TYPE | $I_{CCH}$ (mA) Total with outputs high | | $I_{CCL}$ (mA) Total with outputs low | | $I_{CC}$ (mA) Average per gate (50% duty cycle) |
|---|---|---|---|---|---|
| | TYP | MAX | TYP | MAX | TYP |
| '00 | 4 | 8 | 12 | 22 | 2 |
| '04 | 6 | 12 | 18 | 33 | 2 |
| '10 | 3 | 6 | 9 | 16.5 | 2 |
| '20 | 2 | 4 | 6 | 11 | 2 |
| '30 | 1 | 2 | 3 | 6 | 2 |
| 'H00 | 10 | 16.8 | 26 | 40 | 4.5 |
| 'H04 | 16 | 26 | 40 | 58 | 4.5 |
| 'H10 | 7.5 | 12.6 | 19.5 | 30 | 4.5 |
| 'H20 | 5 | 8.4 | 13 | 20 | 4.5 |
| 'H30 | 2.5 | 4.2 | 6.5 | 10 | 4.5 |
| 'L00 | 0.44 | 0.8 | 1.16 | 2.04 | 0.20 |
| 'L04 | 0.66 | 1.2 | 1.74 | 3.06 | 0.20 |
| 'L10 | 0.33 | 0.6 | 0.87 | 1.53 | 0.20 |
| 'L20 | 0.22 | 0.4 | 0.58 | 1.02 | 0.20 |
| SN54L30 | 0.11 | 0.33 | 0.29 | 0.51 | 0.20 |
| SN74L30 | 0.11 | 0.2 | 0.29 | 0.51 | 0.20 |
| 'LS00 | 0.8 | 1.6 | 2.4 | 4.4 | 0.4 |
| 'LS04 | 1.2 | 2.4 | 3.6 | 6.6 | 0.4 |
| 'LS10 | 0.6 | 1.2 | 1.8 | 3.3 | 0.4 |
| 'LS20 | 0.4 | 0.8 | 1.2 | 2.2 | 0.4 |
| 'LS30 | 0.35 | 0.5 | 0.6 | 1.1 | 0.48 |
| 'S00 | 10 | 16 | 20 | 36 | 3.75 |
| 'S04 | 15 | 24 | 30 | 54 | 3.75 |
| 'S10 | 7.5 | 12 | 15 | 27 | 3.75 |
| 'S20 | 5 | 8 | 10 | 18 | 3.75 |
| 'S30 | 3 | 5 | 5.5 | 10 | 4.25 |
| 'S133 | 3 | 5 | 5.5 | 10 | 4.25 |

¶Maximum values of $I_{CC}$ are over the recommended operating ranges of $V_{CC}$ and $T_A$; typical values are at $V_{CC} = 5$ V, $T_A = 25°$C.

## switching characteristics at $V_{CC} = 5$ V, $T_A = 25°$C

| TYPE | TEST CONDITIONS# | $t_{PLH}$ (ns) Propagation delay time, low-to-high-level output | | | $t_{PHL}$ (ns) Propagation delay time, high-to-low-level output | | |
|---|---|---|---|---|---|---|---|
| | | MIN | TYP | MAX | MIN | TYP | MAX |
| '00, '10 | $C_L = 15$ pF, $R_L = 400\ \Omega$ | | 11 | 22 | | 7 | 15 |
| '04, '20 | | | 12 | 22 | | 8 | 15 |
| '30 | | | 13 | 22 | | 8 | 15 |
| 'H00 | $C_L = 25$ pF, $R_L = 280\ \Omega$ | | 5.9 | 10 | | 6.2 | 10 |
| 'H04 | | | 6 | 10 | | 6.5 | 10 |
| 'H10 | | | 5.9 | 10 | | 6.3 | 10 |
| 'H20 | | | 6 | 10 | | 7 | 10 |
| 'H30 | | | 6.8 | 10 | | 8.9 | 12 |
| 'L00, 'L04, 'L10, L20 | $C_L = 50$ pF, $R_L = 4$ kΩ | | 35 | 60 | | 31 | 60 |
| 'L30 | | | 35 | 60 | | 70 | 100 |
| 'LS00, 'LS04 'LS10, 'LS20 | $C_L = 15$ pF, $R_L = 2$ kΩ | | 9 | 15 | | 10 | 15 |
| 'LS30 | | | 8 | 15 | | 13 | 20 |
| 'S00, 'S04 | $C_L = 15$ pF, $R_L = 280\ \Omega$ | | 3 | 4.5 | | 3 | 5 |
| 'S10, 'S20 | $C_L = 50$ pF, $R_L = 280\ \Omega$ | | 4.5 | | | 5 | |
| 'S30, 'S133 | $C_L = 15$ pF, $R_L = 280\ \Omega$ | | 4 | 6 | | 4.5 | 7 |
| | $C_L = 50$ pF, $R_L = 280\ \Omega$ | | 5.5 | | | 6.5 | |

#Load circuits and voltage waveforms are shown on pages 3-10 and 3-11.

## schematics (each gate)

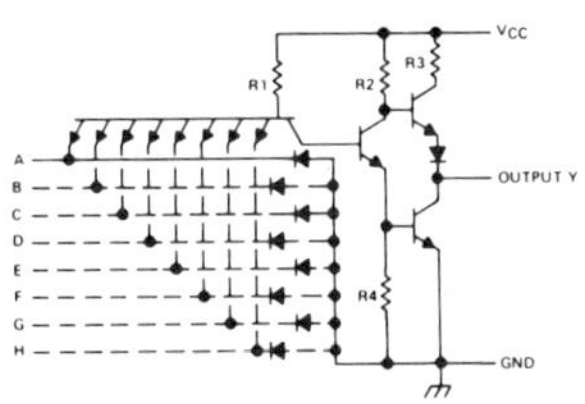


| CIRCUIT | R1 | R2 | R3 | R4 |
|---|---|---|---|---|
| '00, '04, '10, '20, '30 | 4 k | 1.6 k | 130 | 1 k |
| 'L00, 'L04, 'L10, 'L20, 'L30 | 40 k | 20 k | 500 | 12 k |

**'00, '04, '10, '20, '30**
**'L00, 'L04, 'L10, 'L20, 'L30, CIRCUITS**

Input clamp diodes not on SN54L'/SN74L' circuits.

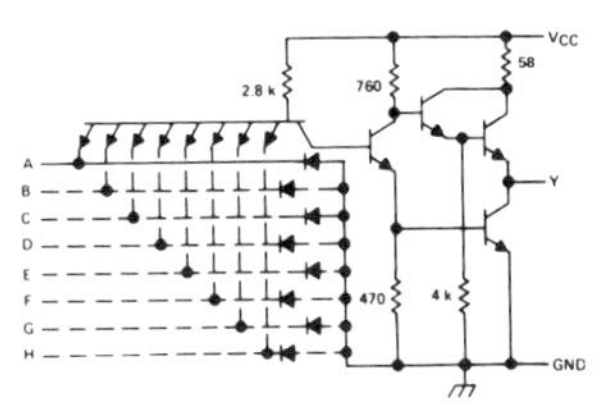


**'H00, 'H04, 'H10, 'H20, 'H30 CIRCUITS**

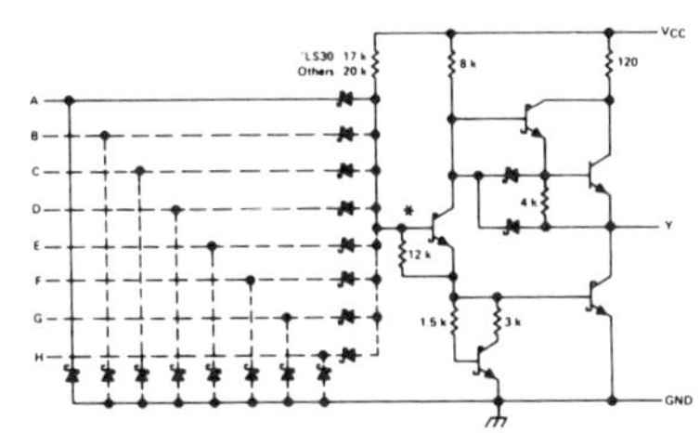


**'LS00, 'LS04, 'LS10, 'LS20, 'LS30 CIRCUITS**

*The 12-kΩ resistor is not on 'LS30.

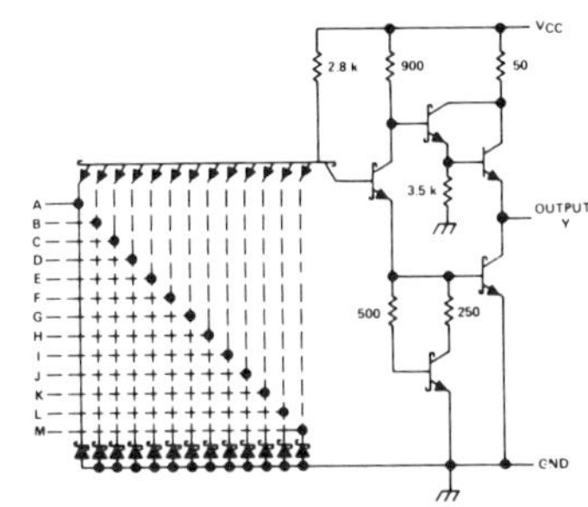


**'S00, 'S04, 'S10, 'S20, 'S30, 'S133 CIRCUITS**

Resistor values shown are nominal and in ohms.

(Motorola), F (Fairchild), and SN (Texas Instruments). The prefix might be modifed. For example, RSN is Texas Instruments' notation for "radiation-hardened circuit," and SNJ is for "joint Army-Navy (JAN)" processed.

Next, and usually of the most interest to the technician, is the series number. The unique circuit description contains the series number. Examples of series numbers are 7410, 7402, 5408, 54L78, 74H10, and 54S10. The 54xx is simply a more rugged version of the 74xx IC. For example, a 7408 is a quad, two-input AND gate guaranteed to operate properly between 0 and 70°C. A 5408 is also a quad, two-input AND gate, but it is guaranteed to operate properly between −55° and +125°C. The 74xx chips are usually used for industrial and consumer applications; military applications usually require 54xx series chips.

Across the top of the data sheet (Table 1-2), we see first the standard 54xx or 74xx series, followed by the 54H/74H series. The 54H/74H series uses smaller-value resistors and a Darlington pair driving the output transistors. The higher speed is accomplished at the expense of increased power dissipation. The low-power series is designated as the 54Lxx or 74Lxx. That series uses larger values of resistors, which reduce power consumption at the expense of longer propagation delays.

The next two series have the highest speed available in the TTL line. They achieve this performance by using a Schottky barrier diode (SBD) connected as a clamp from the base to the collector of each transistor in the gate. This prevents the transistors from saturating and thereby reducing switching time. The 74Sxx, the fastest series of the TTL family, has smaller resistor values and higher power dissipation than the 74LSxx series does.

At this time, TTL families have the highest speed and lowest cost for small-scale and some medium-scale ICs. Unfortunately, they also consume the most power and are more difficult to manufacture on a larger scale (MSI and LSI). The MOS with their lower power consumption and smaller size are becoming better buys, especially if their slower speed can be tolerated.

Table 1-2 is a typical data sheet of terms and electrical parameters. Each parameter is discussed in the order presented on the data sheet. The timing parameters will also be discussed.

The first electrical characteristic parameters are the voltage levels for a valid 1 and 0. $V_{IH}$ is the lowest input voltage that is still accepted as a logic high or 1. $V_{IL}$ is the highest input voltage that is still accepted as a logic low or 0. Refer to Figure 1-12.

Next is the input clamp voltage, $V_{IK}$, which is the maximum voltage that the input clamping diodes will allow the input to go below ground level. The output voltage levels are given next. $V_{OH}$ is the lowest output voltage that is still accepted as a logic high, and $V_{OL}$ is the highest output voltage that is still accepted as a logic low.

The rest of the electrical characteristics are current parameters. Each of these current parameters represent a condition of input current when the input voltage is greater than the logical 1 input voltage specification (2.4 V). This current condition causes the transistor at the input of the gate to operate in the "inverted active mode." That is, the emitter and collector switch roles, and each acts like the other with very reduced $\beta$ or current gain.

$I_{IH}$ is the maximum current that an input line to a digital device will be required to furnish at a proper input high logic level voltage (2.4 V). $I_{IL}$ is the amount of current supplied by the device at an input voltage level of +0.4 V. Remember that the input current of one device must be "sourced" or "sinked" by the output current of a driving device. This is how the fan-out of the device is derived.

The next current specification is the short-circuit output current, which is the amount of current that the device will supply when it is in a logical 1 state and the output is shorted to ground. When this specification is given on a device, it means that the device is short-circuit protected. The output of the device can be connected to ground at any time, and no damage to the device will occur. This protection can be very helpful when troubleshooting. The technician might want to ground the input of a device that can be connected to this device's output. Grounding the input will not damage the driving gate's output if it is short-circuit protected. However, even though the device is protected, the short should not be applied for more than 1 s.

The last current in the table is the supply current $I_{CC}$. For each type of gate, a typical current and a maximum current are given for the outputs, both high $I_{CCH}$ and low $I_{CCL}$. Then the value for the average current per gate is given at 50% duty cycle; that is, the device is on for half the time and off for half the time.

These parameters have been discussed because a technician must understand digital IC electrical parameters to effectively troubleshoot these logic circuits. If the technician does not know for sure what voltage or current *should* be at some point in a circuit, then measuring what *is* there does not help much. Any discrepancies will not be obvious nor helpful in pinpointing trouble.

As stated earlier, the voltage levels and the speed of the device are not ideal. We have discussed at length the voltage. Now, let's consider the timing or speed of these devices.

There is a time delay between the application of the input signal and the response at the output of the gate. This time delay is called *propagation delay*. This delay limits the speed of operation of a complete circuit because it is cumulative through a series of gates that make up the circuit. The table of switching characteristics given on the data sheet (Table 1-2) lists the typical and maximum propagation delay times in nanoseconds for each type/series of gate as the output goes from low to high and then from high to low. The propagation delay waveforms can be measured with an oscilloscope.

# 2

# Digital Troubleshooting Methodology and Approaches

In digital systems, problems that occur become apparent in different ways. In this chapter, you are guided through a logical process of troubleshooting. Some common input and output levels and conditions are presented, and several possible malfunctions are given. Techniques of isolating inputs and outputs are presented along with a discussion of how the isolated points behave.

## What Is Troubleshooting?

In Radio Shack's *Dictionary of Electronics*, the definition of *troubleshooting* is "to look for, locate and eliminate malfunctions in equipment." However, the primary purpose of troubleshooting goes beyond the definition—that purpose being to effectively return the equipment to operational status as quickly and efficiently as possible; that is, to maximize the "uptime" and minimize the "downtime" of the equipment.

### The Competent Troubleshooter

Even the most experienced technician usually begins a troubleshooting project with little idea of the cause of the malfunction. The general "states" that he or she progresses through are illustrated in Figure 2-1. How long the technician stays in each state depends on several factors, not the least of which is skill in troubleshooting.

With the primary purpose of troubleshooting foremost in mind—*put the equipment back into service as quickly and efficiently as possible*—the competent troubleshooter consistently looks for discrepancies between what should be happening and what is happening in the system. To be a skilled troubleshooter, you must approach the project in a certain way:

1. When possible, talk with the equipment's operator; listen carefully for clues as to the source of the trouble. An accurate description of the symptoms by the operator can often save you valuable steps in the troubleshooting process; it can also indicate or eliminate the possibility of operator error.

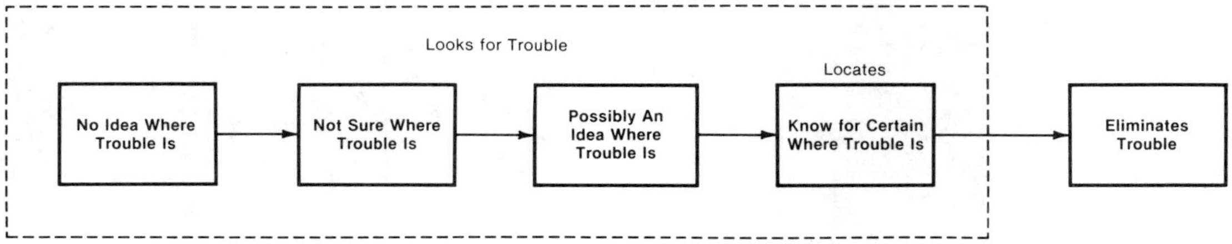


**Figure 2-1.** The "states" that a troubleshooter progresses through to find a malfunction.

2. Attempt the obvious solutions first, such as resetting controls, interlocks, and circuit breakers, and checking for loose connectors and plugs.
3. After eliminating the obvious, begin collecting information by reading meters, counters, and indicator lamps, and/or interpreting oscilloscope waveforms, logic state tables, logic diagrams, and truth tables, using shortcuts when possible. Experience will tell you which source will provide the most data with the least amount of effort and time.
4. Learn to recognize where "locating" the trouble should cease and where "eliminating" the trouble should begin.

Developing competency in the art of troubleshooting requires a balance of correct theory and extensive practice. Both aspects are essential for effectiveness (producing results) and efficiency (producing results with a minimum of effort and time). Without an adequate understanding of the theory involved, you cannot logically approach the problem. Extensive practice, the hands-on aspect of troubleshooting, will enable you to be more selective in your technique—you can begin to select those steps that quickly lead to a solution.

### The Preliminaries of Troubleshooting

Preliminary to any troubleshooting activity must be the verification that a malfunction exists. Often conditions that appear to be serious defects are the result of operator error; that is, the apparent malfunction is improper use of or lack of understanding of the equipment.

Obviously, when troubleshooting any electronic system or piece of equipment, you must know how that system works under normal conditions. You must know the function of all controls, interrupts, and adjustments, how to operate them, and what effect each has on the equipment. If an instruction manual or operator's handbook is available, by all means consult it. Often a readout on the CRT screen provides information of the trouble. A service manual on the equipment, if available, is invaluable. In addition to the schematic and list of parts, some manuals contain a troubleshooting tree. This aid is sometimes in the form of a flowchart that guides the technician through a series of decision situations to locate the trouble.

Many pieces of digital equipment have provisions for a self-test to determine if a trouble is present, thus making the instrument itself a valuable source of troubleshooting information. This test usually occurs when the instrument is initially powered up. Sequencing the equipment through this self-test will assure you of the instrument's integrity and accuracy.

In the absence of a self-test feature, set the equipment's controls for a sequence that should produce known results. A microprocessor system can usually be programmed in a loop to perform a test of each section of the system. Again, be sure that you are familiar with the theory and the operation of the equipment under examination.

## Strategy and Tactics of Troubleshooting

A strategy of troubleshooting is a general approach that works for general situations, as well as for a specific piece of equipment or a particular circuit. Tactics are specific steps that are used to implement strategy. Tactics may vary to fit particular situations. Productive strategy, when properly applied, is similar for most equipment, whether analog or digital, or computers or cars.

## Productive Troubleshooting Strategy

The strategy detailed throughout this chapter will get you from the symptom to the trouble with the least amount of expended time and effort. We will develop a flowchart of an ideal strategy where the specific tactics for applying the strategy to a particular piece of equipment are derived. The flowchart will normally lead you to the trouble in an efficient manner. However, do not hesitate to break out of the flowchart if you can see ahead to the location of the trouble; you will thus decrease downtime.

The goal of productive troubleshooting requires the proficient troubleshooter to return the equipment to service as soon as possible by

- Identifying and verifying the symptoms
- Locating the trouble
- Clearing the malfunction

Your goal is to complete the third step, and you will employ tactics that get you there as quickly as possible.

Thus far, we have examined the more logical avenues of obtaining the information that should lead to the source of trouble and enable you to find and eliminate it. But what if our pursuit has not uncovered the origin of the malfunction? The operator provided no help and the flowchart and self-diagnostics were fruitless. Now what?

When the point of trouble is not located from these direct methods, we must resort to chasing down the source with a systematic, logical procedure. Obviously, this approach is more time-consuming. However, this pursuit is often the only method referred to as "logical troubleshooting." Though they sometimes quickly lead to the source of the problem, the information-gathering tactics discussed previously are usually not considered to be a part of the troubleshooting process.

Most training courses and books on troubleshooting concentrate on the systematic pursuit approach because (1) it is difficult to learn, and (2) even though it is time-consuming and should thus be used as a last resort, it is quite successful in searching out the most elusive malfunctions. Obviously, the most direct sources of information should be employed initially.

## Half-Split Pursuit

Ideally, the efficient pursuit procedure successfully eliminates one-half of the system as the trouble source. We call this technique the *half-split pursuit*. It involves testing the system at its midpoint. By making a test that demonstrates normal operation, you can decide which half contains the malfunction, thus eliminating the other half. Isolating the trouble to approximately half the system is far more effective than either random pursuit or sequentially moving through the system one step at a time. Randomly making tests through the system and indiscriminately replacing parts is the least effective method of locating a trouble. Referred to as the "shotgun" method, its effectiveness depends on luck. The sequential method will eventually lead you to the trouble, but it is inefficient. If you perchance begin your test on the end opposite the trouble, it could take a long time to reach the origin. Note Figure 2-2.

## The Expert

If all the aforementioned methods fail, call an *expert* (a highly trained individual with a wealth of knowledge), if you are fortunate enough to know one. Many equipment distributors and manufacturers can put you in touch with such a person. Speaking with an expert instinctively brings you to the expert's level of understanding; often you solve the problem yourself through this interchange.

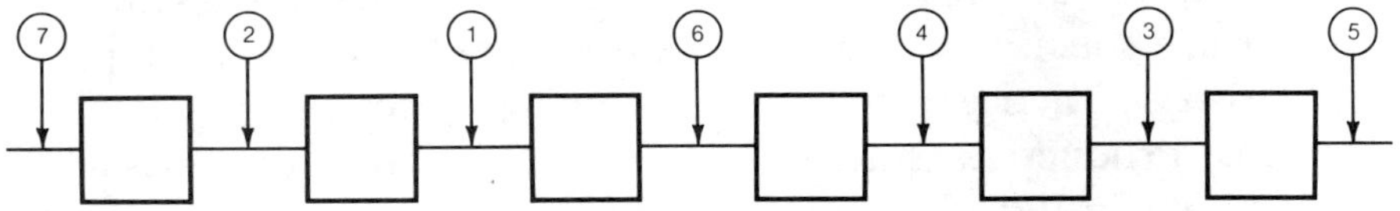


(A) Random tests.

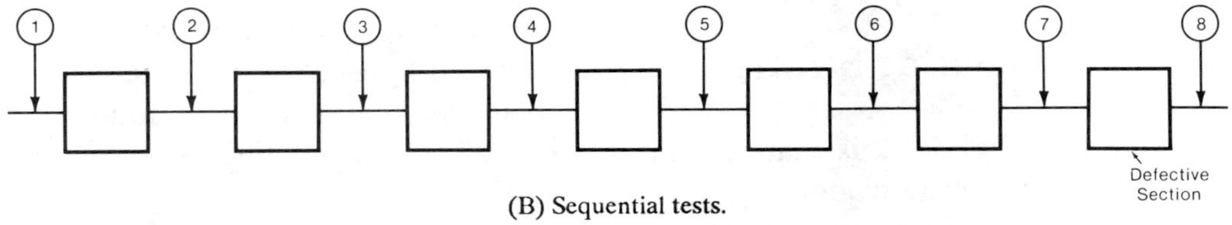


(B) Sequential tests.

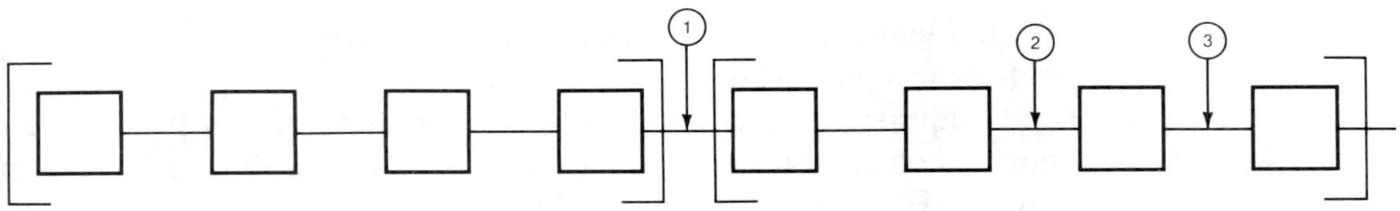


(C) Half-split pursuit.

**Figure 2-2.** Three approaches to troubleshooting.

# A Proposed Pursuit Procedure

The competent technician knows when to turn from an efficient approach to a logically effective method. This section will provide a logical and systematic procedure to effectively find and eliminate malfunctions in digital equipment. With minimum modification, this method can be applied to any electronic, mechanical, or hydraulic equipment. There are four steps to this proposed procedure:

1. Analyze the symptoms of failure.
2. Localize the problem to a complete functional unit or module.
3. Isolate the problem to a logic circuit within the module.
4. Locate the specific problem.

## Analysis of the Symptoms of Failure

When equipment malfunctions, it displays some indication of abnormal performance. Analysis of problem symptoms involves two phases:

1. Symptom recognition—the skill of identifying the abnormal from the normal signs of operation
2. Symptom evaluation—the process of finding a more detailed interpretation or explanation of the abnormal symptoms

Accurate recognition and evaluation leads to a thorough understanding of what the symptoms are and what they indicate.

Do not attack the equipment with a voltmeter or oscilloscope probe until all the symptoms have been carefully examined. Charging into the equipment, measuring waveforms, and taking voltage readings before evaluating the symptoms properly is a waste of time and energy and can well be misleading.

## Localizing a Trouble to a Module

After the symptoms of the malfunction have been properly recognized and evaluated, the next logical step is to localize the trouble to a module. A module is a functioning unit of the system.

On a personal computer system, the modules are typically the computer/keyboard, the

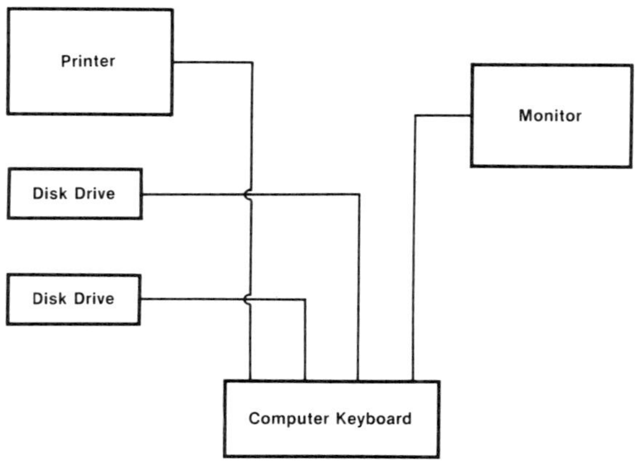


**Figure 2-3.** Block diagram of personal computer modules.

monitor, the disk drive(s), and the printer. The modules of a personal computer are shown in Figure 2-3. They all interact with each other to form the system.

To localize the trouble to a functional unit, use symptom recognition and evaluation to find which units are the probable cause of the malfunction. On systems such as personal computers, where rapid replacement of major modules such as the disk drive is possible, module swapping can quickly isolate the trouble. In the absence of modules available for interchange, a series of well-planned measurements can be helpful. In such measurement plans and system analysis, it is convenient to consider the system as a number of types of paths for signal flow. As in analog systems, these paths will be one or a combination of six basic types, all of which must be operating properly for the system to operate properly.

## Types of Signal Paths

The six basic signal paths are shown in Figure 2-4. Each is an example of a particular digital system or circuit to illustrate an application of the basic type of path.

*Linear Path*—A linear path is a series of circuits arranged so that the output of one circuit is the input of the following circuit. There are no branches or return paths. A group of flip-flops

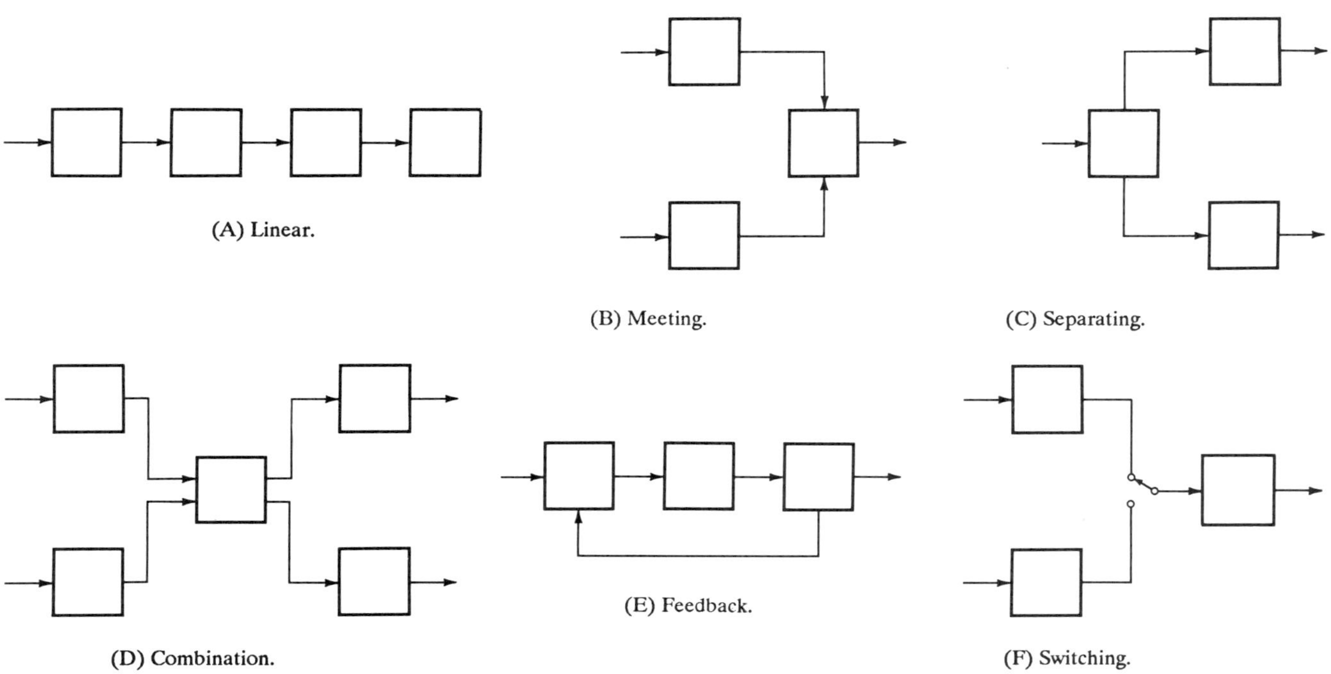

(A) Linear. (B) Meeting. (C) Separating. (D) Combination. (E) Feedback. (F) Switching.

**Figure 2-4.** Six basic signal paths.

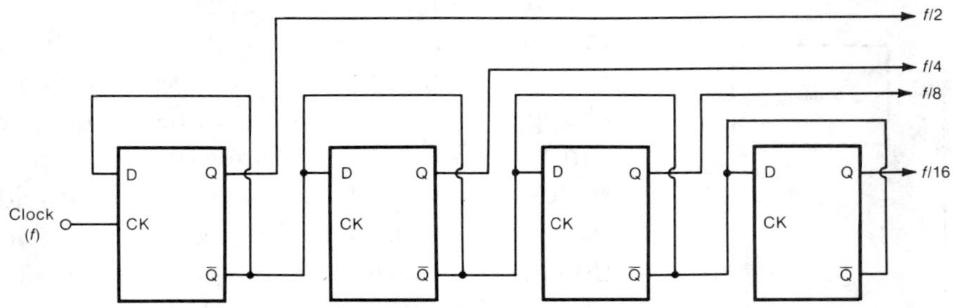


**Figure 2-5.** Linear flip-flops in a row forming a frequency divider.

connected as a frequency divider is one of many places where this type of signal path is found, as shown in Figure 2-5.

*Meeting Path*—In the meeting path, two or more signal paths enter a common point in a circuit. Fan-in of the outputs of logic gates to the input of a single gate is an example of this path type in combinational logic circuits, as shown in Figure 2-6. In data communications, an example of this type of circuit is a multiplex modulation system that combines two or more data channels using time-domain multiplexing.

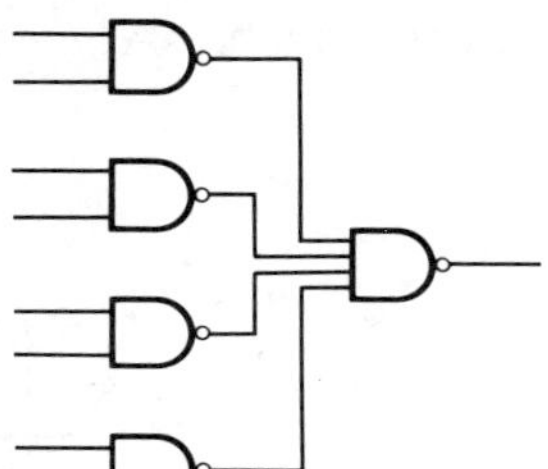

**Figure 2-6.** Meeting or fan-in gates.

*Separating Path*—A separating path is one in which two or more signals leave a common point in the circuit. Fan-out of a logic gate's output to the input of several gates, as shown in Figure 2-7 is a typical application of this type. Again, in data communications, the demodulation of multiplexed channels is another example of this type of separating signal path.

*Meeting/Separating Path*—In a meeting/separating path, the circuit or module has multiple inputs and multiple outputs. A decoder, such as the 7447A shown in Figure 2-8, is a common example of this type. It is a binary-coded-decimal (BCD) decoder with 4 inputs and 7 outputs.

*Feedback Path*—The feedback path is similar to the linear path except that there is a signal path from one point in the path to a point preceding it in the signal flow sequence. A modulus-to-ripple counter such as that in Figure 2-9 is an example of a feedback path circuit. Here, however, the feedback is to the CLEAR of each flip-flop.

*Switching Path*—The switching path contains a selector switch, either electrical or electronic, that provides a different input or output path for each

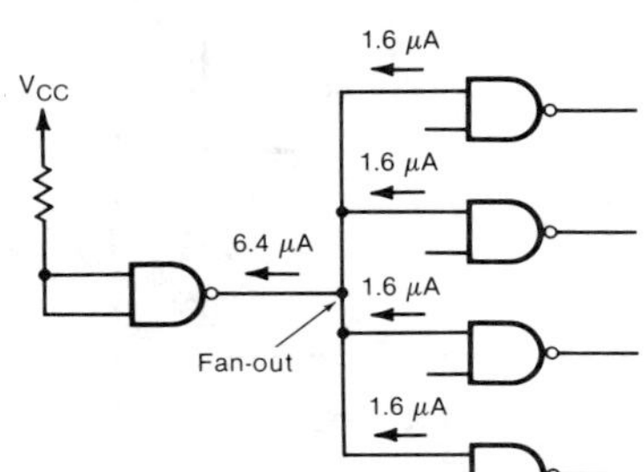


**Figure 2-7.** Fan-out to four gates showing origins of all circuits.

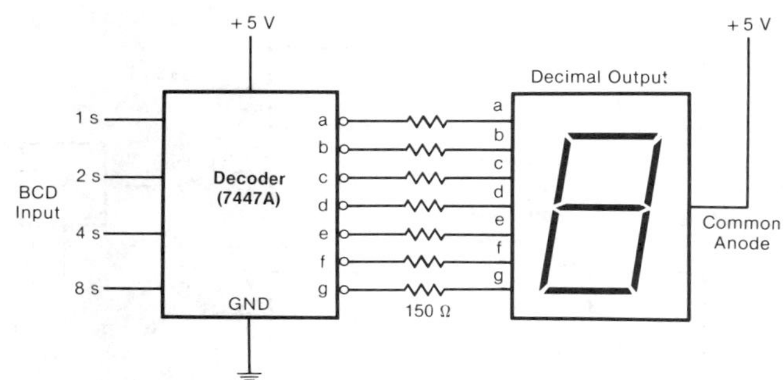


**Figure 2-8.** Wiring of a decoder and 7-segment LED display.

**Figure 2-9.** Modulus-10 counter.

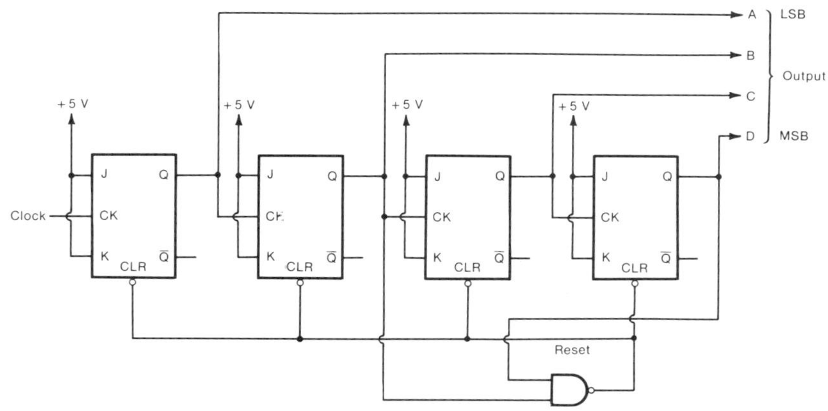


switch selection. A computer system with two disk drives is an example of the switching path, both in the read and the write modes.

### Half-Split Pursuit

The half-split pursuit method of chasing a trouble can be specifically applied to the problem of localizing the trouble to a module or subsystem. This technique is best suited when the signal path is linear. Unless the symptoms point definitely to one module, circuit, or circuit group as the probable trouble source, the most logical place to make the first test is at a convenient location halfway between the point known to be good and the point known to be in trouble. It is very difficult to test a system exactly at the midpoint of a section to be checked, and it is also unnecessary. The objective is to test at a point each time that will eliminate a large portion of the system from suspicion.

Some systems consist of large circuit boards containing many components. Changing boards on these systems can quickly isolate the trouble to a particular board. Because it can be done with little time and effort, this technique is quite useful, even though the particular board is not at the midpoint of the system.

## Isolating the Trouble to a Logic Circuit

After analyzing the symptoms and localizing the trouble to a module, you will have the information needed to isolate the trouble to a probable faulty unit. The first two steps require a minimum of testing and measuring. The isolation step involves extensive use of test equipment, first to isolate the fault to a group of circuits within the module, and then to isolate the faulty circuit from the group. The isolation process for a linear path follows the same logical pursuit procedure described earlier—that of continuously narrowing down the possibilities by making observations and measurements and making logical decisions from that information.

### Isolating a Trouble Involving a Separating Path or a Meeting Path

When a signal path is a separating path or a meeting path, a variation in the procedure is required. Separating and meeting paths are not as simple and easily recognized as linear paths. Refer to the definitions of these paths. Before the trouble in these paths can be isolated, the paths themselves must be isolated. A simple example of a separating and meeting path circuit is shown in Figure 2-10.

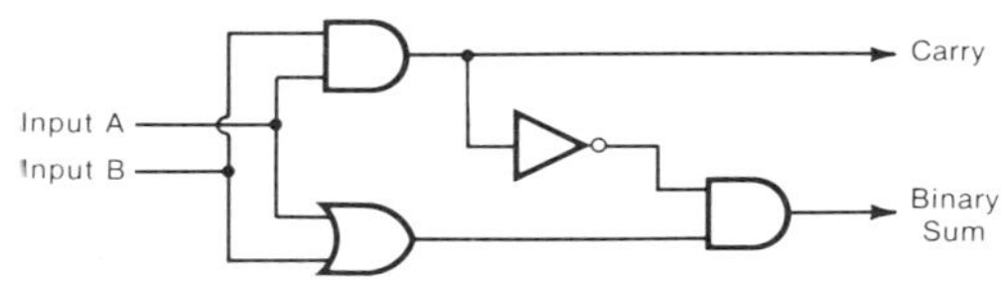


**Figure 2-10.** Separating and meeting path circuit.

Figure 2-10 is a half-adder circuit group that consists of replaceable gates and is part of a computing circuit. The circuit failed to compute the sum during a test run, and the trouble was localized to the half-adder circuit, which was then replaced by swapping the small printed circuit board (PCB) holding the circuit group. The card is now connected to an appropriate power supply, pulses are supplied by a pulser, and the various points are monitored with a logic probe or oscilloscope (see Chapter 6).

To test the half-adder, inject a 1 pulse at the digit A of the OR gate and check for a 1 pulse at the sum output as well as a 0 (no pulse) at the carry output. A proper response indicates that both the OR gate and the B AND gate are functioning normally. Confirm this by injecting a 1 at digit B input, and again check for a 1 at the sum output and a 0 at the carry output.

If the responses from the sum or carry outputs are not proper, the OR gate or the B AND gate or both are defective. To localize the problem further, inject a pulse at either digit A or digit B input of the OR gate and check for an output from the OR gate. If there is no pulse output from the OR gate or if the output is abnormal, the problem is in the OR gate. If the output is normal, the problem is likely in the B AND gate.

Now, to complete the test of the half-adder circuit, inject simultaneous pulses at both digit A and digit B inputs and check for a 1 (pulse) at the carry output and a 0 (no pulse) at the sum output. If the response is proper, all the gates can be considered as functioning normally. If not, the nature of the response can be analyzed to isolate the trouble to a gate.

What if all tests are proper? Now what? Carefully examine and clean the edge connector on the board. Then reinsert the board in the system and recheck the system. If the board works out of the system, then it should work in the system. The trouble can probably be traced to poor or dirty connecting terminals, which might have been found right away by wiggling the connection, thus saving all that testing time.

## Locating a Specific Trouble

The final step of the pursuit procedure is to locate the specific trouble to an individual component. Many troubles (not all) are cleared by replacing sealed modules or integrated circuits (ICs). The use of replaceable modules often minimizes the number of steps required to localize and isolate the trouble, but it is still necessary to check branch circuits to components and controls outside the module. Controls and adjustments as well as DIP (dual in-line package) switches for programming functions are good examples, because they are not always located within the sealed module, but are instead connected to terminals of an IC or a PCB. For example a microprocessor chip was changed three times before a defective capacitor on one of its pins was discovered. This locate step will enable you to find the specific cause of trouble, then remove the trouble and return the equipment to service.

## A Visual Inspection

The competent troubleshooter is always on the lookout for something suspicious. A visual inspection is particularly important after the trouble has been isolated to a circuit. of course, all of the senses should be used, but sight and a large measure of common sense are useful tools in locating shorts, opens, and charred components.

Once the defective component is identified, tests should be made before replacing it, or the new component may likewise be destroyed. For example, a short may burn a resistor so that it is charred. If you replace the resistor without removing the short, the new resistor will also be burned out.

A review of all information obtained thus far—symptom analysis, tests and measurements, visual inspection—should verify that the defective component that has been located is the exclusive cause of the trouble. If the suspected component can produce all of the normal and abnormal symptoms that were observed, it is logical to assume that it is the sole cause of the trouble. If not, use your knowledge of electronics, digital logic, and the particular piece of faulty equipment to search further.

When a component fails, the cause is usually an operating condition that exceeds the maximum ratings of the component. A component can also simply become defective (see discussion on static discharge in Chapter 10). The operating condition that causes the component failure can be temporary. A solid-state component can be destroyed instantaneously, for they are not as forgiving as vacuum tubes. Occasionally the cause of failure is the component's basic design.

## The Checkout

In the strictest sense, repair (that is, replacement of the defective component, clearing a short, completing an open) is not part of the troubleshooting procedure. It is, however, necessary to the total purpose and will be discussed in detail in Chapter 10.

After correcting the malfunction, perform an operational checkout to verify that the equipment has no further faults. Never assume that simply because a defective component has been located and replaced, the equipment will automatically operate normally again. In practical troubleshooting, assume nothing and prove everything.

# Organizing a Troubleshooting Procedure

A systematic plan of troubleshooting can make the project more efficient. A good way of organizing a procedure can be borrowed from the digital computer programming field. Both the programmer and the troubleshooter must organize the task and then develop a logical plan to follow. A programmer usually develops a flowchart to aid in reasoning through the problem. Let's apply the concept of the flowchart to the organization of a troubleshooting procedure.

## Flowcharts

A flowchart is a diagram that shows the logical flow or sequence of procedural steps of a task. Every task that is undertaken, such as troubleshooting, must be identified in terms of actions that must be performed in the proper sequence—a logical plan to follow. Much effort can be saved by the use of a flowchart, and it provides a greater assurance that the complete procedure has been accomplished. In flowcharts, symbols define the general type of operation that is to take place at each step in the procedure. These symbols are shown in Figure 2-11.

(A) Entrance and exit. (B) Do something. (C) Decision. (D) Connector.

**Figure 2-11.** Flowchart symbols.

The first symbol, Entrance and Exit, is used only at the beginning and end of a flowchart. The words written inside the symbol signify the meaning to be assigned to it. Usually the word Start signifies the beginning, and the word Stop or Exit signifies the end.

The rectangular shape in Figure 2-11B is used to imply that something is to be done in that particular step in the flowchart procedure. Typical labels inside these rectangle symbols might be Verify Symptoms or Make Final Checks.

The diamond-shaped symbol in Figure 2-11C signifies that a decision must be made. In troubleshooting—digital or analog, electronic, hydraulic, or mechanical—many decisions must be made. Every time a measurement is performed or information obtained, the troubleshooter must determine whether that measurement is within tolerance. In digital troubleshooting, measurement decisions are based on the simple criteria of whether the measured parameter is a valid high or a valid low, or whether the high or low is present for the required time interval. Examples of statements that are found in decision boxes are Operator Induced Trouble?, Trouble Cleared?, Power Supply Voltage Normal?, Gate U6 Input Level Low?.

The circle in Figure 2-11D is used for connecting parts of the flowchart; no action is performed. The flowchart is limited by the size of the page. When the edge of the paper is reached and additional flowchart is needed, this connector symbol is used. An identifying letter is drawn inside the circle and at the point on that page or another page to signify the restarting at that particular point.

Flowchart symbols are connected with straight lines that have arrows at the receiving end of the line to show the direction of the logical process. By using combinations of the flowchart symbols and straight lines, you can develop an easily visualized, logically organized procedure for troubleshooting. A general strategy of troubleshooting in a flowchart form is shown in Figure 2-12.

# Analyzing and Correcting Digital Output and Input Troubles

Problems that occur in digital systems usually come to our attention because that system's output fails to function properly. The output of a digital system may appear in various forms: a cathode ray tube (CRT), a printer, a display of light-emitting diodes (LEDs), or a liquid crystal display (LCD). The output may control a motor or other electrical device through a digital-to-analog device. All of these types of outputs and displays are driven by the outputs of hardware logic circuits. The troubleshooter will be testing the outputs and inputs of these circuits following a systematic troubleshooting procedure similar to that in Figure 2-13.

## TTL Devices

The majority component on a PCB is the IC chip. ICs are usually mounted on PCBs in one of two ways: they are either soldered directly in the

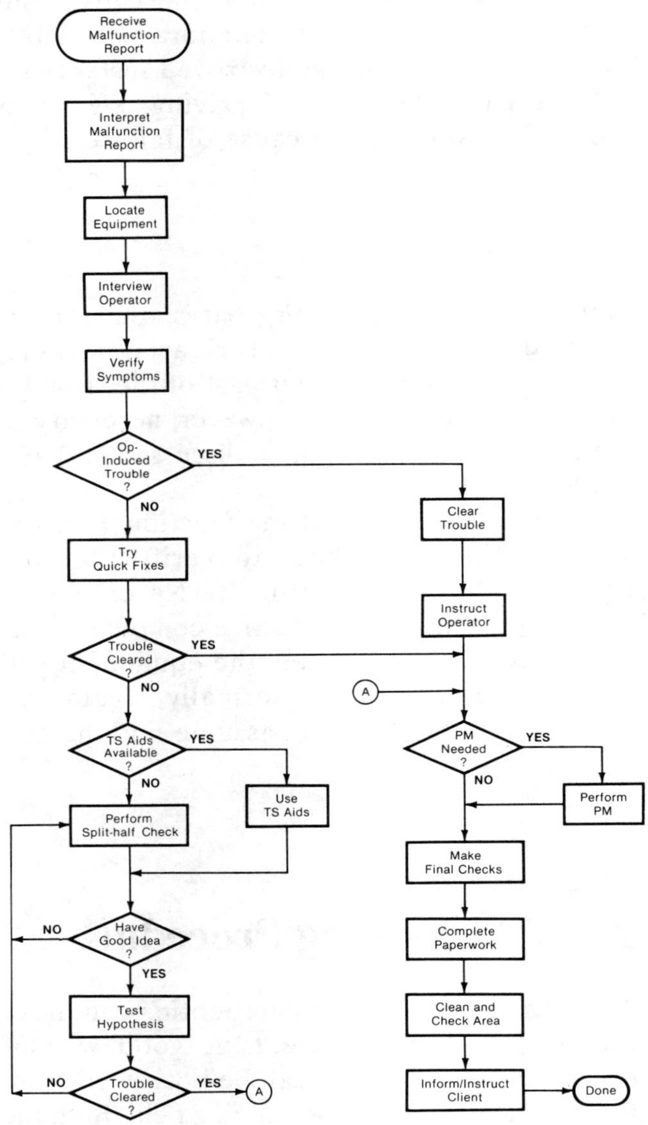


**Figure 2-12.** A strategy of troubleshooting using a flowchart.

Bookmark

*HOWARD W. SAMS & COMPANY*

DEAR VALUED CUSTOMER:

Howard W. Sams & Company is dedicated to bringing you timely and authoritative books for your personal and professional library. Our goal is to provide you with excellent technical books written by the most qualified authors. You can assist us in this endeavor by checking the box next to your particular areas of interest.

We appreciate your comments and will use the information to provide you with a more comprehensive selection of titles.

Thank you,

Vice President, Book Publishing
Howard W. Sams & Company

COMPUTER TITLES:

Hardware
- ☐ Apple I40 ☐ Macintosh I01
- ☐ Commodore I10
- ☐ IBM & Compatibles I14

Business Applications
- ☐ Word Processing J01
- ☐ Data Base J04
- ☐ Spreadsheets J02

Operating Systems
- ☐ MS-DOS K05 ☐ OS/2 K10
- ☐ CP/M K01 ☐ UNIX K03

Programming Languages
- ☐ C L03 ☐ Pascal L05
- ☐ Prolog L12 ☐ Assembly L01
- ☐ BASIC L02 ☐ HyperTalk L14

Troubleshooting & Repair
- ☐ Computers S05
- ☐ Peripherals S10

Other
- ☐ Communications/Networking M03
- ☐ AI/Expert Systems T18

ELECTRONICS TITLES:
- ☐ Amateur Radio T01
- ☐ Audio T03
- ☐ Basic Electronics T20
- ☐ Basic Electricity T21
- ☐ Electronics Design T12
- ☐ Electronics Projects T04
- ☐ Satellites T09
- ☐ Instrumentation T05
- ☐ Digital Electronics T11

Troubleshooting & Repair
- ☐ Audio S11 ☐ Television S04
- ☐ VCR S01 ☐ Compact Disc S02
- ☐ Automotive S06
- ☐ Microwave Oven S03

Other interests or comments: ____________________

____________________

____________________

Name ____________________

Title ____________________

Company ____________________

Address ____________________

City ____________________

State/Zip ____________________

Daytime Telephone No. ____________________

*A Division of Macmillan, Inc.*
*4300 West 62nd Street*
*Indianapolis, Indiana 46268*

**22571**

Place
Postage
Here

HOWARD W. SAMS & COMPANY

Dept. DM
4300 West 62nd Street
Indianapolis, IN 46268

board or inserted in a socket that is soldered directly into the board. Figure 2-14 shows a PCB with both soldered-in chips and chips mounted in sockets. Though more expensive to manufacture, the socket-type IC mounts are much easier to troubleshoot.

A digital system may consist of just one circuit board or many circuit boards. Circuit boards must be connected to peripheral controls, displays, and other parts of the system. When several PCBs are used in a system, as in Figure 2-15, they are usually mounted in a card rack, with edge connectors making the connections between the boards. The edge connector allows power and signal connections to and from PCBs without soldering wires to the boards.

The boards are easily removed and reinstalled in the connectors. However, this is also a source of trouble. A poor connection because of a dirty, corroded, or loose-fitting edge connector is a common source of trouble. It is sometimes intermittent. Removing and reinserting the board temporarily clears the trouble. In this case, the edge connector contacts should be cleaned. A common pencil eraser makes a very good abrasive to clean contacts, especially the male edge connectors of PCBs.

The trouble may be erractically intermittent and may respond to slightly twisting the PCB. Sometimes you can pull or push a wire in a plug, jack, or edge connector and cause the intermittent trouble to occur. The plugs and their mating connectors may make and break connections that can usually be cured by respringing the contacts so that they fit tightly

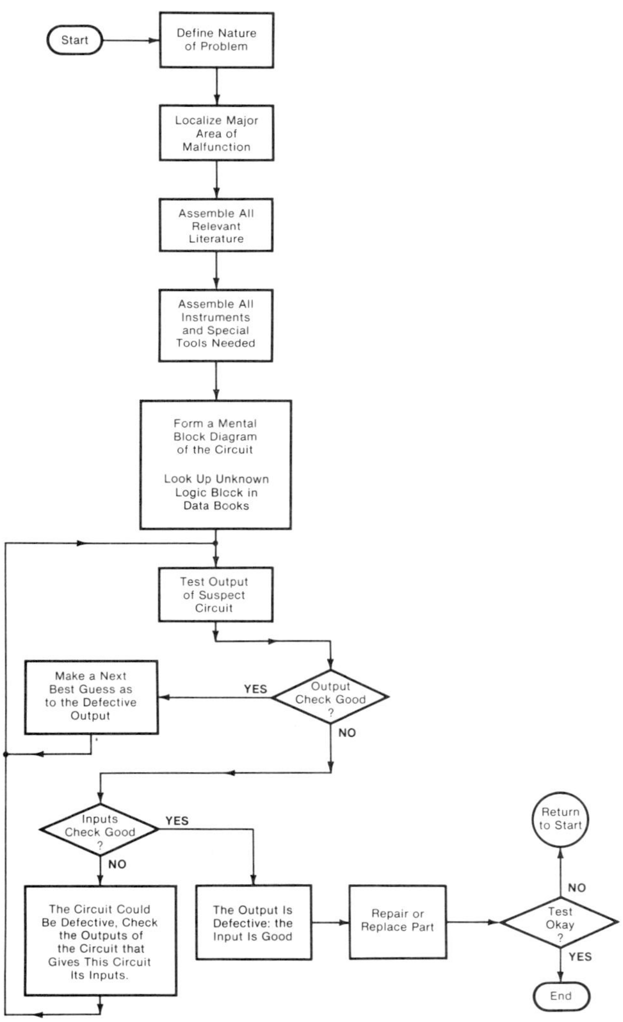


**Figure 2-13.** A troubleshooting flowchart showing a systematic method of testing inputs and outputs.

**Figure 2-14.** Mounted integrated circuit chips.

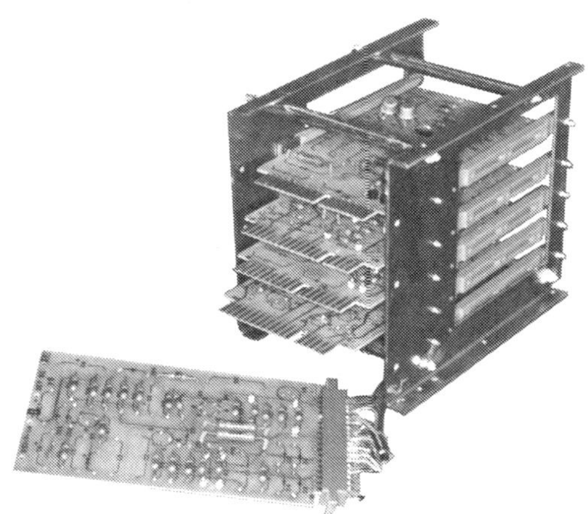

**Figure 2-15.** Several PCBs in a card rack.

together again. There are many different types of connectors, so the type of contact between them also varies greatly. A relay adjustment tool has been quite handy for respringing connector contacts. Sometimes you have to replace the connector for the affected card or board.

In a card rack, the boards are so close together that it is nearly impossible to put a test probe on the circuit board. An extender card or board extender, shown in Figure 2-16, allows the PCB to be brought out to where it may be accessed easily for testing. This extender is merely a one-to-one direct-wired extension for the edge connector.

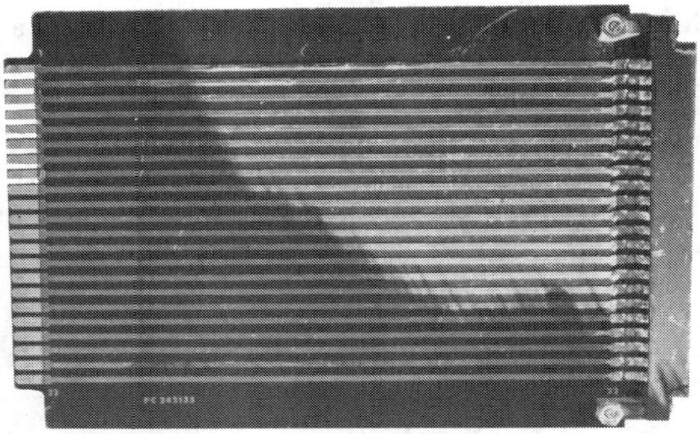

**Figure 2-16.** An extender card.

## Checking Logic Output Levels

In digital systems, the output is easy to test because it has only the two states—logical 1 and logical 0. The ideal voltage levels for a 1 and a 0 and the fact that a practical 1 and 0 had a range of values that are valid were discussed in Chapter 1. Suppose the output voltage were 1.5 V. Is this a logical 1 or a logical 0? The digital troubleshooter frequently encounters this problem. It should be clear from our discussion of "sink" and "source" currents in Chapter 1 that if an output level is not a valid logical 1 or a logical 0, it can be experiencing an excess load. That is, you must know what state the output is supposed to be in, because excessive load in either state will cause the gate's output to be in the nonvalid region.

Now let's look at the invalid output voltage levels and their probable causes. For convenience Figure 1-12 is repeated as Figure 2-17. This shows that the valid range for a logical 0 is from +0.06 to +0.8 V. The output levels should be examined carefully. The oscilloscope is best suited for making this measurement because it not only can resolve voltage values to a few millivolts, but it can measure both static (constant) and dynamic (changing, pulsing) conditions. The proper use of an oscilloscope is discussed in Chapter 6.

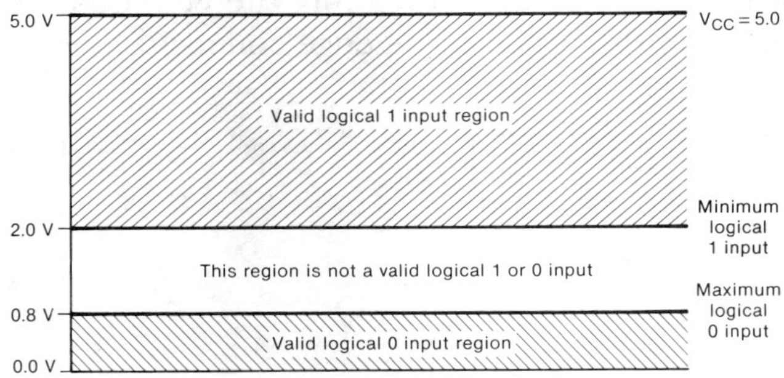


**Figure 2-17.** Boundaries for valid input voltages equal to logical 1 and logical 0.

Two conditions represent the most common output levels found in the logical 0 state. First, let's consider the output of two or more open-collector gates tied together as shown in Figure 2-18. In Figure 2-18A, the gates A and B each have a high (logical 1) and a low (logical 0) on their inputs. This should give a high (logical 1) on the common output. The output measures a low instead. Gate B has a shorted output, which pulls the common output for gate A low also. In fact, careful examination of the level will probably indicate an output of less than +0.07 V, which you should check for a possible short to ground. Since the gates have a passive pull-up (open-collector) output structure, as shown in Figure 2-18B, if either gate A's or gate B's output goes low, the common output goes low. In this case, gate B's output is abnormally low because of its shorted output. However, either gate A or gate B could cause identical conditions with shorted outputs. Also, a short at ground on the PCB traces anywhere on the output node would cause

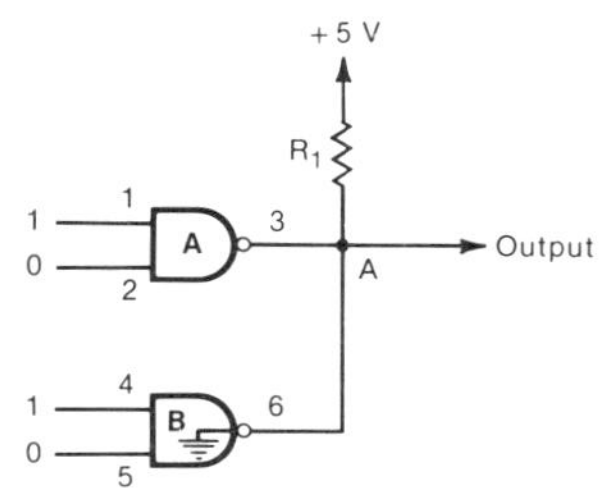


(A) Two open-collector NAND gates tied together (one has a shorted output).

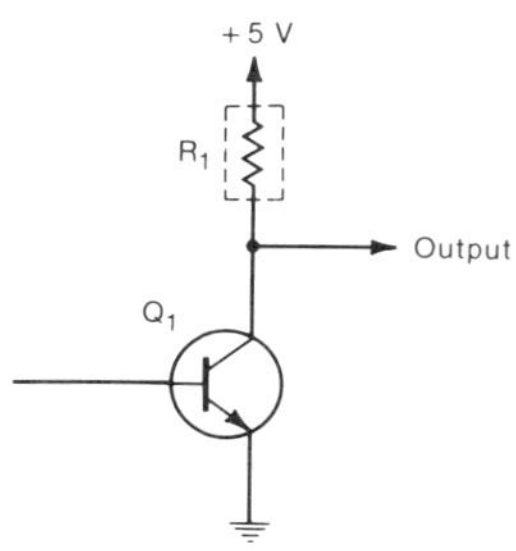


(B) Passive pull-up (open-collector output structure of a TTL gate).

**Figure 2-18.** Output levels.

identical symptom conditions. To find the short, you must isolate it. The outputs of gates A and B can be lifted one at a time so that when the short is removed from the node, the level of the output will go high. The techniques of lifting the outputs from the node (isolating the outputs) will be discussed shortly.

A special instrument for tracing shorts (current drains) is the current tracer, which is discussed in Chapter 6. Another method of finding a point shorted to ground uses the same idea but makes use of Ohm's law and voltage-drop difference on a trace that is carrying a heavy current (one shorted to ground). The voltage drop along the trace is proportional to the current through it. The sensitivity and resolution of the voltmeter used to measure the drop along the trace must be high enough. Either an oscilloscope or a digital multimeter is usually sufficient to measure both sensitivity and resolution. You can measure the voltage drop using the output node (the bottom of $R_1$) as a reference for the scope or digital voltmeter (DVM). Then as the probe is moved along the traces, the voltage drop will increase in the negative (−) polarity as the probe gets nearer to the shorted point to ground. The voltage drop will be only a few millivolts, so the voltmeter, oscilloscope, or DVM must be sufficiently sensitive.

When examining a logical 1 voltage on an active pull-up gate, note that the device may have several common output voltage level conditions. First, the valid level is between 2.0 and 5.0 V (the valid level range), as shown in Figure 2-19. A normal voltage range is between about +2.5 and +3.5 V. If the output voltage level is approximately equal to $V_{CC}$, the output is probably shorted to $V_{CC}$ either internally to the device or externally on the PCB. If the output of the device is less than +2.0 V but is not a valid logical 0, the device is usually being overloaded. Excessive current is being supplied by its output to some point that is probably shorted or nearly shorted to ground. This situation occurs when the device is being required to source more current than it is capable of sourcing. Its output, therefore, drops below the valid high (1) voltage level.

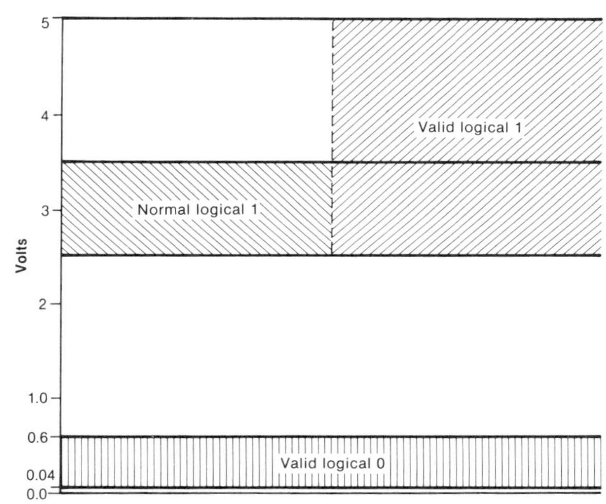


**Figure 2-19.** Valid output voltage levels for TTL gates.

A pulse waveform on the output of an active pull-up device that is attempting to go alternately high and low is shown in Figure 2-20. The high is a valid voltage level, but the low is in the invalid range. This indicates that two device outputs are probably shorted together somewhere, with the second device trying to maintain a logical high, which is interfering with the output voltage waveform on the first device.

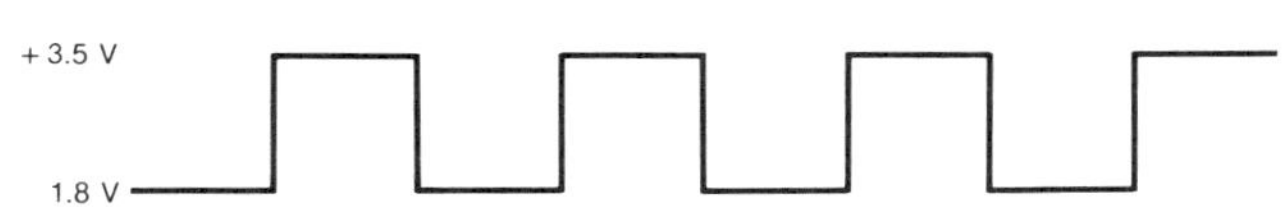


**Figure 2-20.** A pulse waveform on the output of a device whose output is shorted to another.

## Isolating an Output

The output voltage we have discussed indicates a probable defect at an output. However, that output must be isolated to determine whether the trouble is in the output of that device or the input of the next device that is being driven. The technique used to isolate an output of a suspected defective device depends on several factors:

- Is the device in a socket or soldered directly in the PCB?
- Does the output go to another point on the same circuit board or to another point off that board? Is that connection by a wire or plug from board to board or is it by an edge connector?
- Is the output of the suspect device active pull-up or passive pull-up (open collector)?

First, let's consider isolating an output pin from the PCB that the chip is mounted on. If the IC is socket-mounted, then the isolation job is simple. Take the IC chip out of the socket. Being careful not to break the output pin, bend it up and reinsert the chip in the socket. The isolated output of a NAND gate is shown in Figure 2-21.

**Figure 2-21.** Isolating the output pin of socketed IC chip.

If the chip is soldered directly on the PCB, the techniques are not as simple as with socketed IC chips. One technique to isolate the output of a soldered-in chip is to cut the pin of the device as close to the PCB as possible, as shown in Figure 2-22. Then bend the pin up so that it is not in contact with the PCB, as shown in the figure. Now that the device's output is isolated, it can be further tested. Now bend the pin back in contact with the board and carefully resolder it.

**Figure 2-22.** Isolating the output pin of a soldered-in IC chip.

Another technique, which does not require cutting the device pin, involves unsoldering the pin and removing all the solder by using a solder sucker or wicking. Then free the pin by rotating it back and forth in the hole with a pair of needle-nose pliers. Slide a small piece of "spaghetti" insulating tubing over the pin and through the hole. The output pin is now isolated and may be tested (this testing is described later) and easily resoldered.

If the output goes to another PCB or to some point off its own board and if that connection is by a wire that can be easily disconnected, mark the wire and where it goes with masking tape. Then unsolder it and proceed to test the isolated output. If an edge connector is used to interconnect the PCB, find the pin of the edge connector to which the output of the device is connected. Place a piece of masking tape or cellophane tape over the edge connector pin on the circuit board. Trim off the excess tape around the pin with a hobby knife or razor blade. Be sure that the tape goes over the edge of the board so that it will not be pushed off and expose the connector to be isolated when the board is reinserted in the connector socket. If the edge connector is single-sided, fit the tape around the board to the other side. The output is now isolated and can be tested.

### Checking the Output

After the output has been isolated, there are some considerations that are helpful and necessary in troubleshooting and testing it. First, you must know if the output of the device is active or passive pull-up. If the device is passive pull-up, then disconnect the pull-up (load) resistor from the output of the device. Therefore attach a temporary replacement pull-up resistor from the output pin to $V_{CC}$.

Now, measure the voltage level at the output and make a decision from the information obtained. If the output seemed to be shorted to ground before it was isolated and after it was isolated it goes to a valid high or low, the device is probably good and the short on the board is elsewhere. If the output level was neither a valid high nor low before isolation and goes to a valid logical state or toggles (switches) from one state to the other after isolation, then the output is most likely being overloaded. Though it is possible that the device is unable to properly sink or source its rated current value, it is most likely a good device.

From the test that you have just made, you can decide if the device is good or bad, and if the trouble is in the output path. Further testing of the device is treated in Chapter 6, when test equipment is discussed with applications.

## Checking Logic Input Levels

The inputs of the devices are tested in much the same way that the outputs are tested. The isolation techniques are identical. However, a few things should be pointed out. First a shorted-to-ground input will not allow the voltage level to toggle to the valid 0 and 1 states. If the input is shorted to ground, the voltage level will measure 0.0 V and produce an excessive load on the device that is driving it. If the input is shorted to $V_{CC}$, it will measure a continuous +5 V, which will also load the driving device when it tries to go to a low.

The most common condition is when the input is open, either in the IC or on the PCB. If the gate is open inside the chip, it will appear that the input is toggling properly from 0 to 1, but the voltage is correct only outside the chip. *Note: An open TTL input will behave electrically as if a logical 1 is applied to the input.* The voltage reading on a TTL input that is "floating" (disconnected from its driving source) will measure around +1.5 V, which is an invalid logical level.

### An Open Circuit on the PC Board

To find an open on the PCB, you should make several incremental voltage meausurements. If you find two places on the same node that measure differently, there has to be an open between them. Here are two very helpful troubleshooting rules:

1. A voltage cannot be developed across a short.
2. A current cannot be developed through an open.

Several places on the PCB give this type of trouble. Here is a brief list of some common places:

- A small hairline crack on the PCB trace; a magnifier/light is a must for finding this trouble.
- If the IC socket insertion spring tension gets weak and one or more of the device's pins do not make good electrical contact, there will be an open.
- The device could have a pin bent under the chip instead of being inserted in the socket hole.

- If the IC chip is not socket-mounted, the pin may be in the correct hole but not soldered to the PCB.
- An IC chip may have been inadvertently reinserted in the socket backward.
- If the IC socket has been wire-wrapped for device interconnection, a wire may have broken or there may have been a miswired connection on the prototype.

It is a good idea to add to this list as your experience builds. The techniques dicussed here fit many troubleshooting problems found in industry. They have been proven repeatedly to be the best in getting the equipment back up and in service as fast as possible.

# 3

# Microcomputer Fundamentals

Most digital troubleshooting and repair jobs require some basic knowledge to effectively localize the problem. Thus this chapter presents concepts and operations of the microcomputer, including the 8086 microprocessor. Architecture, software, and interfacing techniques are covered in sufficient depth to allow troubleshooting and repair of a system using a microprocessor as its base. The bus interface unit (BIU) is described along with control signals, such as memory read or write and input/output (I/O) read or write. Registers and memory are discussed briefly. We also compare the 8-bit and the 16-bit microprocessors.

## Introduction and History

Most introductions to any computer system begin with a history of computing. Where to begin with this history is quite arbitrary. We might start with our early ancestors using their fingers (digits) to relate the concept of two sabertooth tigers or three mammoths. Or we could begin several centuries BC with the abacus, an early digital calculator. In 1642, Blaise Pascal, at the age of 19, invented a desk calculator to add long columns of numbers for his father, who was a tax collector. The unique feature of his calculator was its ability to carry a digit to the next place when a gear passed the zero location.

In 1833, Charles Babbage proposed a machine design that would perform a complete sequence of operations as specified by punched cards. He sufficiently impressed the English government to provide funds to initiate construction, but his ideas were too far advanced for the mechanical technology of his time. The mathematical basis for today's digital designs can be traced to the work published by George Boole in 1854.

International Business Machines (IBM) was incorporated from a company formed by Herman Hollerith in 1880. Its original purpose was to manufacture punched-card tabulation equipment for the U.S. Census Bureau. In 1937, Claude Shannon showed how to use symbolic logic in digital circuit design, and Howard Aiken, at Harvard, proposed to IBM a design for a computer that would be constructed from over 3,000 electromechanical relays. It was called the Mark I and was completed in 1944.

The next big step toward today's computer was the construction of ENIAC, begun in 1943 by S. P. Eckert and J. W. Mauchly at the University of Pennsylvania. It had no moving parts and, instead of relays, relied on 18,000 vacuum tubes to accomplish its task.

In 1945, John von Neumann, at Princeton University, started a computer development project in conjunction with the University of Pennsylvania. The major concept that came from this project was that of storing program instructions and data in a memory so that instructions could be modified in the course of solving a problem.

The general-purpose, stored-program computer had become commercially available to the business community by the early 1950s. From then to now, the emphasis, in addition to expanded application and programming, has been to make the computer less expensive, smaller, and more reliable.

The major point of this brief history is that the concept of the microprocessor is not new. Today's microprocessor bears a remarkably strong architectural resemblance to the early computer. In fact, what is inside a modern microprocessor is essentially a smaller, more cost-effective version of the stored program that John von Neumann conceptualized and developed in 1945. Only the manufacturing techniques are truly new. The improved manufacturing techniques have reduced price, size, and operation time requirements while greatly increasing performance and reliability.

# Introduction to Microcomputers

## What a Microcomputer Is and What It Is Not

From the first electronic digital computer in the 1940s to modern revolutionary microcomputers, as well as the full-size mainframe computers, basic computer operation has changed very little. New technologies, such as transistors and integrated circuits (ICs), have reduced the size and cost of the computer. However, digital computers still process the 0s and 1s using the same basic logic operations as did their predecessors. A single IC, using large-scale integration (LSI), can now perform all the functions of a computer. Thousands of such circuits can be manufactured at a cost not much more than that of simple circuits.

A human being can do the same thing as a computer, but the computer performs the operation with much greater speed and accuracy. Also, the computer works untiringly. It performs its task a single step at a time, though its speed of a microsecond or less per step compensates for this apparent inefficiency. However, unlike its human counterpart, the computer has to be given a complete set of instructions that tell it exactly what to do at each operational step. This set of instructions, called a *program*, is stored in the computer's memory. These instructions are then taken one at a time from the memory, and the computer performs each step in sequence.

The hierarchy of computers begins with the giant mainframe computer and progresses down through the minicomputer to the microcomputer. Most of the principles discussed in this chapter are applicable to most computers in the hierarchy. However, we will concentrate on the smallest and newest member of the computer family—the microcomputer. It generally consists of several IC chips, including a microprocessor chip, memory chips, and I/O interface chips.

## Basic Computer System Organization

As is true with other systems, a computer system can be broken down into its various components. This breakdown is shown in Figure 3-1.

All of the adding, subtracting, shifting, and logical operations are performed by the arithmetic and logic unit (ALU). The memory of the computer system contains two types of informa-

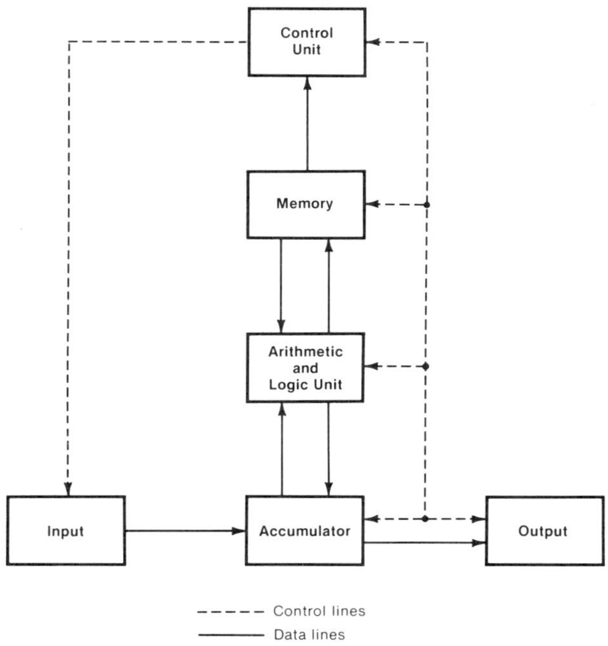


**Figure 3-1.** Computer system organization.

tion: instructions and data. The accumulator is a one-word, temporary memory used to hold data that is operated on by the ALU. The instructions are binary numbers that are passed (fetched) from memory to the control unit. The control unit decodes each instruction and sends out signals to all other system components, which causes gates to open and close, counters to increment and decrement, and registers to shift to execute the particular instruction. In order to communicate with the computer system, input and output devices are required. These devices, called *peripherals*, include keyboards, magnetic tapes and disks, control sensors, cathode ray tube (CRT) video monitors or liquid crystal displays (LCDs), and printers.

The groups of lines or wires that connect the parts of the system together are called the control bus, the data bus, and the address bus. Standard practice is to refer to the individual lines of a bus by numbers. The first is numbered 0, the least significant bit (LSB), and on up to 7 for an 8-bit bus and 15 for a 16-bit bus. For example, the address bus labels are A0–A15.

The control section is grouped with the ALU in a computer system discussion. The resulting combination is termed the central processing unit (CPU). The CPU performs the functions of fetching, decoding, and executing the instructions of operations to be performed on the data. Along with the I/O sections and the system buses, the CPU gives the outside world a means of communicating with the computer. All of this requires timing and control, which is accomplished by the system clock and control bus. A computer is, therefore, an electronic machine capable of quickly performing complex calculations by means of stored instructions and data.

## Hardware, Software, and Firmware

To understand how instructions are given to the computer, we must first understand memory types. Generally, the types can be classed as temporary (volatile) and permanent (nonvolatile). The temporary memory of a computer can be written to and is referred to as *random-access memory* (RAM). Random access means that each memory location is equally accessible. A more descriptive name for RAM is read/write (R/W) memory. The nonvolatile memory is called *read-only memory* (ROM). Although it is accessed randomly also, this type of memory cannot be written to. The instructions in the memory of a computer system are usually in one of three forms: hardware, software, and firmware.

*Hardware* refers to the physical elements from which the computer is fabricated—the electronic, mechanical, and magnetic components that physically compose the computer system. Hardware can be connected in such a way that the computer performs a series of operations to complete a given task. This system is referred to as *hard-wired*. The program of instructions is connected to do a specific task. If a different task is desired, the wiring must be changed; that is, the hardware must be reprogrammed by changing the hard-wiring.

Most computer systems require a software program of instructions to be stored into memory; the hardware is as useless without this software as the software is without hardware.

*Software* is the instruction package that controls the operation of the computer step by step from beginning to end. Software programs are usually first written on paper and then entered on the keyboard, thus transferring the instructions to the computer's memory. Programs stored in RAM are referred to as software.

Programs stored in ROM can be changed only by changing the ROM. Since these programs have properties of hardware and software, they are termed *firmware*. Firmware, then, refers to programs or software instructions permanently stored in a ROM.

## Control of Electronic Computer Systems

The microcomputer consists of several sections that must work together as a system. These sections include the CPU, memory units, and various I/O interface units. The clock and associated timing and logic circuit provide the necessary control signals to coordinate the activities within the microcomputer. A typical microcomputer control system is shown in Figure 3-2.

The microprocessor's internal states as well as external circuits connected to the system bus are changed synchronously by the pulses produced by the clock generator circuit, the bus control logic, and/or the microprocessor itself. The speed with which the microcomputer operates depends on the pulse rate of the clock. Every microprocessor has a maximum rate, which places an upper bound on the computer's operational speed. Sometimes the computer requires more than one train of clock pulses. Each train is called a *phase*. A clock that produces two such pulse trains is referred to as a *two-phase clock*. This is typical for microprocessors such as the Intel 8080 or the Motorola 6800. However, three, four, or more phases may be required for some computer systems.

The control bus is one of the most important buses in the system because it controls the memory and the I/O devices, and synchronizes the operation of the total computer system. Each type of microprocessor has a slightly different control bus configuration. The most important, and the most common to all microprocessors, of the control bus signals are the READ and WRITE signals. These signals synchronize the basic functions of the memory and the I/O circuits. In addition, there are special inputs and outputs, such as interrupt control, direct memory access control, ready, reset, and various other processor-dependent control signals. Direct memory access (DMA) is a method whereby an external device goes directly into the memory and either extracts or stores information. DMA is used for data transfer at extremely high speeds.

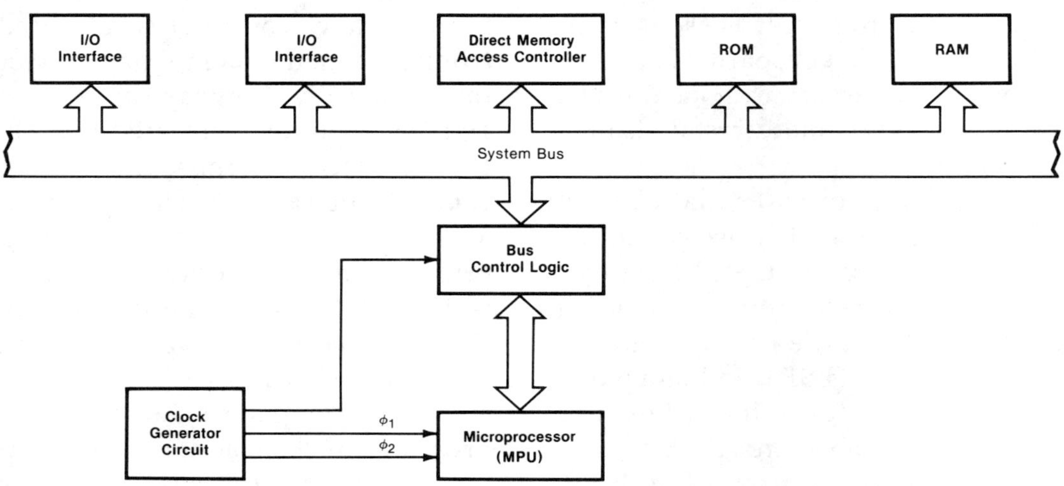


**Figure 3-2.** Typical microcomputer control system.

The operation shown in Figure 3-1 depicted the layout of most bus-structured digital computer systems. The flow of information between itself, memory, and I/O equipment is accomplished by a central unit. In a mainframe and a minicomputer, this central unit is called the CPU. As depicted in Figure 3-2, this function is accomplished by a microprocessing unit (MPU) in a microcomputer.

## The Microprocessor—Brain of the Microcomputer

Three main functions are shared by all MPUs: data transfer, arithmetic and logic operations, and decision making. The most important is data transfer, because the MPU spends most of its time transferring data to and from memory. Besides fetching instructions and data from memory and storing data in memory, the MPU transfers data to and from the I/O and moves it around in the microprocessor itself. This is done by a register array referred to as *scratch pad* memory.

The next most common function of the MPU is arithmetic and logic operations. By far the most often used feature is addition. By complementing a number and adding it to a second number, the MPU can perform subtraction. Though some newer microprocessors are capable of multiplication, it is usually accomplished by successive addition. All of the basic logic functions may be performed by the microprocessor, although the AND function is the most commonly used. AND is used mostly to mask or clear to 0 a portion of a binary number. The OR function can be used to set bits to 1s. The Exclusive-OR function is used to invert or give the complements of bits. The MPU sometimes has operations that test the bits in its accumulator.

The function of the MPU that is least used is that of making a decision based upon some numerical test. For example, a number in an accumulator can be tested to see if it is 0 or if it is negative. The result of this test can be used to control the execution sequence of the program. These program instructions are in the form of a conditional jump or conditional branch. This function neither processes any data with the ALU nor transfers data to or from memory. This function is a powerful tool that allows the MPU to execute the same specified code again and again until the status of the test condition is met, which allows the jump or branch instruction to be executed.

### An Overview of Microprocessor Architecture

The term *architecture* is used to describe the style of construction of the computer under discussion—in this case the microprocessor. It also describes the arrangement and size of the registers, bus configuration, or structure. Figure 3-3 shows the architecture of a hypothetical microprocessor with all of its prevalent features. Later in this chapter, we will look at the specific architecture of several popular microprocessors as we compare their operation and specifications. The internal architecture of the hypothetical microprocessor will now be used to describe the functional components and their interaction.

A microprocessor contains at least the blocks, control unit, control memory, ALU, and working register, and, either on the same chip or on a support chip, additional memory, I/O interfaces, and the clock and timing control circuits.

### The Clock and Timing

As stated previously, the trains of pulses from the clock circuitry coordinate the activity of the microcomputer. The clock circuitry is a pulse generator that outputs one or more pulse trains that have the same frequency but different phases. Each phase during a clock cycle initiates a different action to be taken while an instruction is being executed. An instruction cycle is the

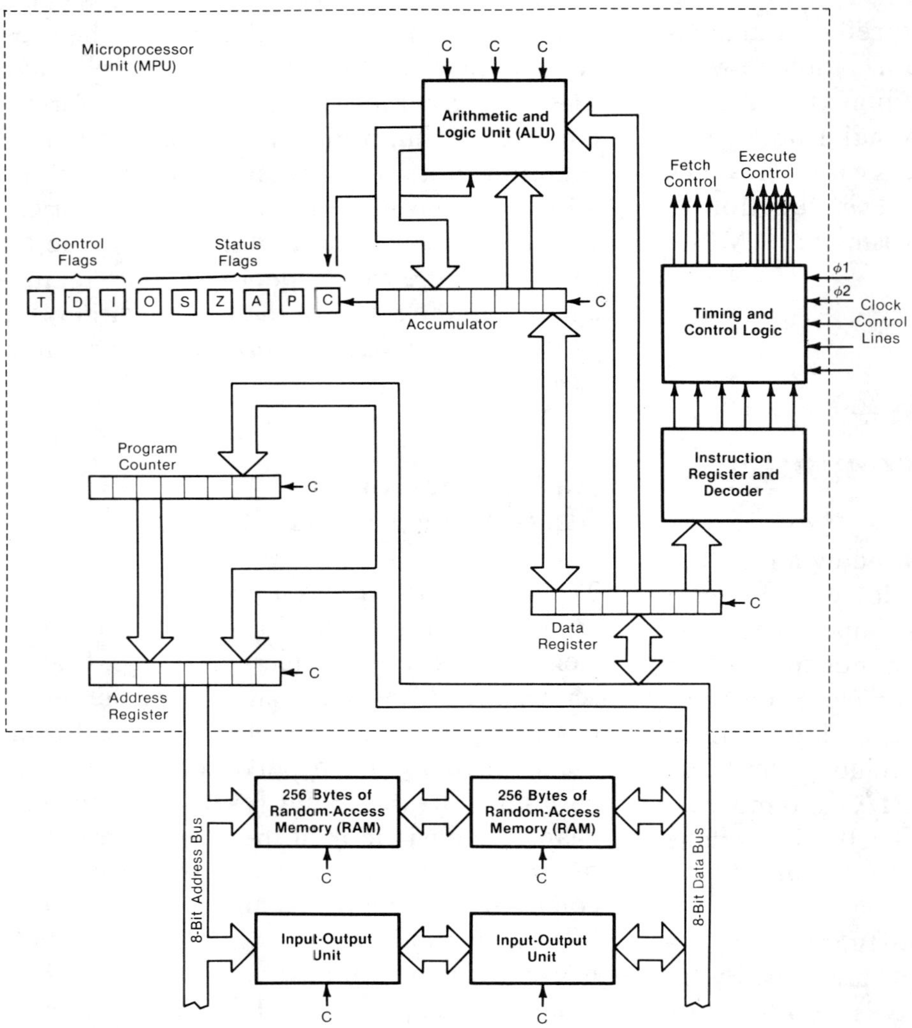


**Figure 3-3.** Architecture of a hypothetical microprocessor.

combination of these actions taken during any particular instruction. Some instructions take more actions than others; thus the instruction cycles are further divided into machine cycles and states. A *machine cycle* is the interval of time that memory or an I/O device is accessed. A *state* consists of the actions that occur during the machine cycle. For example, the first machine cycle of an instruction would be to retrieve the first byte of instruction code from memory. This is called the *fetch cycle* and may consist of four or five states, depending on what other actions must be taken in addition to fetching the instruction byte. Most machine cycles consist of three states to access memory. For example, to read from memory requires the following states:

- Send address to memory
- Memory looks up requested information
- Information is transmitted from memory to CPU

Other operations can be broken down similarly. Any instruction that accesses memory needs at least one state for memory to look up information. The memory quite often needs more than a clock cycle (state) to complete its task, especially if very slow memory is being accessed. Many microprocessors have a "wait state" for the remainder of the system to allow time for the memory to complete its operation.

The hypothetical microprocessor of Figure 3-3 shows a two-phase clock. The two signals, phase 1 and phase 2, are generated on the microprocessor chip and are used both internally and externally to synchronize every action of the CPU. The two clock signals are the same frequency but are phase-shifted with respect to each other.

In Figure 3-4, the clock signal is detailed during a READ operation along with the related signals on the system buses (control, address, and data). This READ operation occurs during every fetch cycle of an instruction. The fetch cycle may do any of the following operations:

- Fetch the operational code from memory
- Fetch the address of an instruction or data from memory
- Fetch the actual data from memory

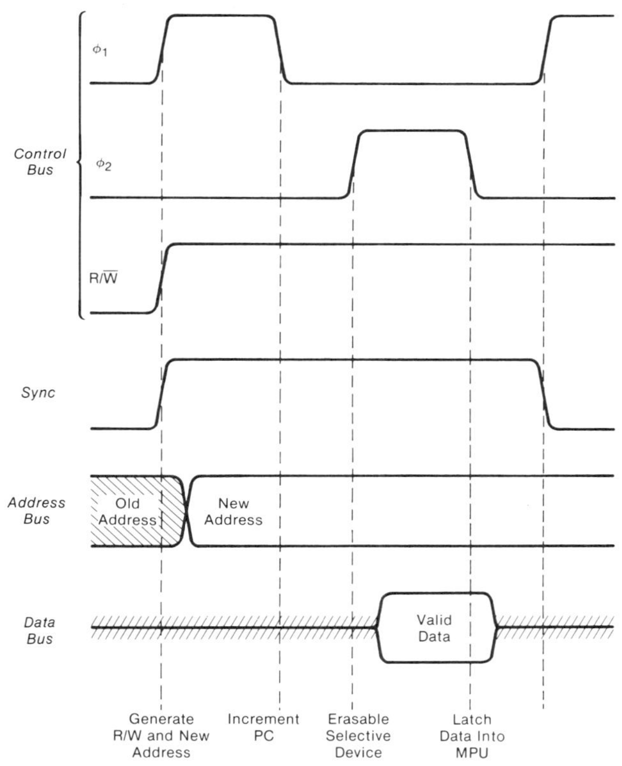


**Figure 3-4.** Clock signals and bus activity during a read to memory operation.

Examination of this timing diagram shows that everything is referenced to the clock signals, with each rising and falling edge of each clock phase producing an operation. At the rise of the phase 1 clock pulse, several things should be noted. The R/$\overline{W}$ control line goes from low to high, which puts it in the READ mode. The address bus will have the "new" address placed on it, which determines the location from which the MPU is to read information (memory or I/O). The double lines on the address bus indicate that the individual lines of the bus have a combination of 1s and 0s on them to make up the address. The data bus will be in the high-impedance state until phase 2 goes high. Also, no device tied to the data bus will be enabled until phase 2 is in its high state. When the phase 1 clock signal goes low, the program counter (PC) is incremented in the control unit, but no bus activity is noted.

The rising edge of phase 2 is the next interval of bus activity. While phase 2 is high, the selected memory or I/O device will put its data on the data bus, and the individual data lines of the bus will go high or low according to the data applied through their three-state data buffers. All of the unselected memory or I/O devices will be unaffected, because their three-state data buffers will remain in the high-impedance state.

The last state of the READ cycle is entered when the phase 2 clock signal goes low. In this state, the microprocessor takes the information from the data bus, puts it in one of the internal registers (most likely the instruction register or the accumulator) and latches it there. The memory or IC device must be capable of applying valid data on the data bus before the falling edge of phase 2.

The address bus activity and the data bus activity both show a change that is delayed a short time after the clock phase transition that produced them. This delay is due to propagation delays that occur in the microprocessor's internal circuitry and is the main reason that memory and I/O devices have an operating speed that is compatible with the clock frequency. Notice that the data bus returns to the high-impedance state after the falling edge of phase 2.

During this READ operation, only three signals of the control bus are used: phases 1 and 2 and the R/W lines. The control bus has other signals, some of which are sent to other parts of the microcomputer. Some are signals that come to the MPU from these other elements. Each MPU has its own unique set of I/O control bus signals. These signals will be described in more detail later in this chapter, when we compare various microprocessors, and in the next chapter, when we discuss families of chips and peripherals. Now, let's discuss some control bus signals.

The RESET (RES) signal is used to get the MPU started or restarted at the time it is powered-up or anytime thereafter. It does several things (each micoprocessor being unique), but commonly it initializes the MPU's internal circuits and sets the PC to the address of the first instruction to be executed, which is stored in the ROM.

The READY (RDY) input control line and the WAIT output control line are used to add extra clock periods (states) to the basic machine cycle in order to access slow memories or I/O devices.

The SYNC line provides this output signal, which identifies the start of each memory access cycle. Notice in Figure 3-4 that this signal went high on the first rising edge of the phase 1 signal and low on the next rising edge of the same phase 1 signal. This function is called DATA TERMINAL READY (DT/R) in some of the modern MPUs.

The DATA ENABLE (DEN) is the signal that caused the three-state devices on the data bus to go in and out of the high-impedance state.

The interrupt requests are input signals that interrupt the execution of the current program and cause the MPU to jump to a special program called the *interrupt service routine*. After this program is completed, the MPU returns to the main program and continues. There are two basic types of interrupt signals: maskable (IRQ) and nonmaskable (NMI). The proper signal on the NMI line will always interrupt the MPU, whereas the IRQ line can be disabled so that the MPU will not respond to its interrupt request.

These are only samples of the most commonly encountered control bus signals. Others will be discussed as they are needed.

## The Registers

A *register* is a small temporary memory used to hold data or an address during an instruction cycle or between instruction cycles. It has been noted that the most common operation that takes place inside the MPU is the transfer of binary information from one register to another. As we will see, the major differences between various MPUs are the types and number of registers and how they are utilized.

The most used register in the MPU is the accumulator. When the ALU is performing arithmetic or logical operations, it performs a dual function. Before the ALU operation, it holds one of the operands. After the operation, it holds the result. There are several instructions for every microprocessor that utilizes the accumulator. A "load accumulator" instruction causes the contents of some memory location to be transferred to the accumulator; a "store accumulator" instruction causes the binary contents of the accumulator to be stored in some location in the main memory.

The accumulator may sometimes be included in a group called *data registers*. Data registers are used for temporary storage of data that are going onto or off the data bus or that are being stored in memory. It may also be used to hold an instruction that is being decoded. Data registers are usually used during program execution for temporary storage of frequently used intermediate results. The advantage of storing these data in internal registers instead of memory is that they can be accessed much faster.

The PC contains the address of the memory location with the next instruction. The instruction cycle begins with the controller placing the contents of the PC on the address bus. The first word of instruction is then fetched from memory and put in the data register. The controller then increments the contents of the PC so that the next instruction cycle will fetch the next sequential instruction from memory. The instruction word is transferred from the data register to the instruction register (IR), which holds the instruction until it can be decoded.

When the operation code word has been decoded by the instruction decoder, it then signals the control logic to generate the proper sequence of control and timing signals to complete the execution of the indicated instruction.

Index registers are used for addressing. The contents of the index register are added to the memory address that an instruction would otherwise use. The sum is then the actual address of the data or the "effective" address. The use of this register will be clearer when we discuss index addressing later in this chapter.

The stack pointer (SP) is a special kind of register that allows the programmer to use a section of RAM as a last-in–first-out (LIFO) memory called the *stack*.

Condition code registers, sometimes called status registers, hold individual bit indicators called *flags*, which represent the status of a particular MPU condition. They are occasionally used to indicate the condition of external serial inputs or outputs. The state of each flag can be checked by the microprocessor. The result of a particular flag is the basis for the microprocessor's decision making. The number and types of flags vary with different microprocessors. For example:

1. The Motorola 6800 microprocessor has six condition codes or status flags: negative (N), zero (Z), overflow (V), carry (C), interrupt (I), and half-carry (H). These six flags make up bits 0 through 5 of an 8-bit register. Bits 6 and 7 of the condition code register are not used and are always set to 1.
2. The condition code or flag register in the 8086 microprocessor by Intel is a 16-bit register with only nine of its bits being used for flags. Six bits are used for status flags, three for control flags. The status flags are called carry (CF), parity (PF), auxiliary carry (AF), zero (ZF), sign (SF), and overflow (OF). These status flags indicate conditions that are produced at the completion of execution of instructions in the ALU. The control flags are the string direction (DF), the interrupt (IF), and the trap (TF). These control flags differ from the status flags (sometimes called conditional flags) by the way they get set or reset. Status flags are set or reset by the execution unit (EU) on the basis of the results of some arithmetic or logical operation; control flags are deliberately set or reset with specific instructions in the program. Seven of the 16 bits in the flag register are unused or undefined at this time. The locations of the nine used flags in the 8086 register are shown in Figure 3-5. The operation of these flags and how they are used with software will be discussed later in this chapter along with the use of the other registers.

## Arithmetic/Logic Unit

One of the most important circuits in the microprocessor is the ALU. Its purpose is to perform an arithmetic or logic operation on the data words that are delivered to it. The ALU has two main inputs. One comes from a register called the accumulator, and the other comes from the data register on the hypothetical microprocessor of Figure 3-3. The controller sets the operation of the ALU as directed by the instruction decoder. The ALU receives two 8-bit or 16-bit binary words, generally from the data register and the accumulator or other working registers. These binary words are called *operands* and may be added, subtracted, or logically combined in some manner. The result is then stored back in the accumulator. The output of the ALU is also used to send a set of status signals to the condition code or processor status register to set or clear various flag bits. For example, if the operation on the two operands by the ALU is such that the result is 0, a signal is sent to the Z flag to set it to the 1 state. If the addition of the two binary numbers produces a carry out of bit 7 (the most significant bit of an 8-bit microprocessor), a signal is sent to set the C flag to a 1.

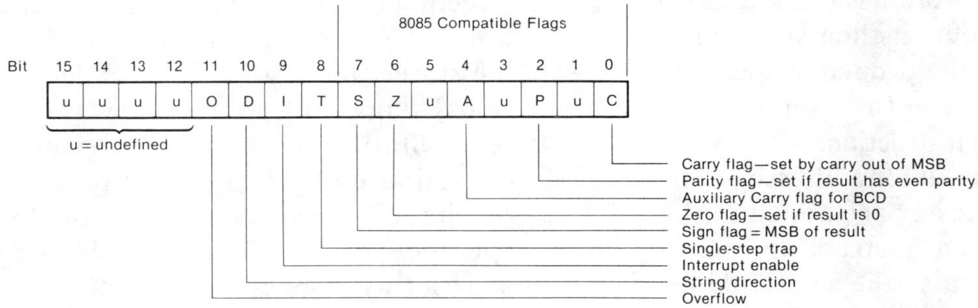


**Figure 3-5.** The 8086 flag register format.

Notice that the ALU in Figure 3-3 also has an input from the C flag. The C flag can be set or cleared by instructions in some microprocessors (6502). The C flag then takes part in various shift and rotate instructions.

Most modern microprocessors have ALUs that can perform a wide variety of arithmetic and logical operations. The specific functions that the ALU performs are determined by control signals generated from the control logic under program instructions.

## Memories

The memory capabilities are what make digital computers adaptable and versatile compared to analog systems. The digital computer is a stored–program-type system that tells the computer what to do under every condition.

We have discussed registers, a type of memory used for the internal operation of the digital computer. Each register is capable of storing a single word or byte of information.

The external, sometimes called main, memory is used to store the instructions of the program and the data on which they operate. Each word, and usually each byte, has an address that the MPU uses to access it. Main memories, as shown on the hypothetical microcomputer of Figure 3-3, are semiconductor memories called ROM and RAM. Semiconductor memories are well suited for the high-speed memories of a microcomputer. However, their cost per bit of storage prohibits their use as mass storage devices.

Mass storage refers to memory that is external to the computer. Such memory has the capacity to store millions of bits of data without the need for constant electrical power to maintain its ability to hold data. Popular mass storage devices are magnetic tape and magnetic disks. They are much slower than semiconductor memories, but are popular for mass storage because they are less expensive per bit. We will look at tape and disk storage when we consider I/O interfaces.

Another type of mass memory catalog is the magnetic bubble memory (MBM), a semiconductor device that uses magnetic principles to store millions of bits on one chip. Though it is a semiconductor type, it is very slow and therefore cannot be used as an internal memory to the computer.

Every memory system requires several different types of I/O lines that select the address in memory that is being accessed, that select either a READ or a WRITE operation to be performed, and that transfer the information to and from memory on the data bus. When several memory modules are combined to form a larger memory, the memory system has some means to enable (or disable) all or part of the memory so that it either will or will not respond to the address bus signals and the READ and WRITE commands.

### Semiconductor Memories

The most common types of semiconductor memories used with microprocessors are ROMs and RAMs. RAMs are more properly referred to as random-access read/write memories.

ROMs are the most common memories in microprocessor systems. They are usually programmed by the manufacturer as part of their manufacturing process. They are much less expensive in large quantities than RAMs are because they require only circuitry for reading the memory addresses. They are about 10 times more dense than RAMs and easier to interface to the microprocessor system. ROMs are programmed by the manufacturer and cannot have their contents changed. They can only be read from. However, some ROMs, called programmable read-only memories (PROMs), can be programmed by the user with the proper equipment. Another type of ROM, called erasable programmable read-only memories (EPROMs), can be erased with ultraviolet light and reprogrammed many times. Still another type of ROM is the electrically alterable programmable read-only memory (EAPROM). The EAPROM is programmed in the same way as the PROM—by applying an address to several of its pins and then applying a write voltage to a write pin. However, an EAPROM is not permanent. When used with properly designed R/W circuitry, the EAPROM may be reprogrammed with a computer. However, the EPROM must be programmed with a special piece of equipment called an EPROM programmer.

The entire family of ROMs is used to store data and information that is not changed during the operation of the system. A major use for ROMs is the storage of control programs in the microcomputer. Since the family of ROMs is nonvolatile, these programs are not lost when the microprocessor is turned off. When the microcomputer is turned on, it can immediately begin executing the program stored in the ROM, PROM, EPROM, or EAPROM. ROMs have one, two, or more control inputs that have to be activated in order to read information on the data outputs. They are called by different names, such as chip select (CS), chip enable (CE), or output enable (OE). Of course, there is no R/W input because a ROM cannot be written to except under special conditions (e.g., EAPROMs).

Many types of memory can be described as having random access, which means that any memory address location can be as easily accessed as any other. When the term RAM is used with semiconductor memories, it specifically means that the memory may be capable of READ and WRITE. The RAM is volatile; that is, it loses its stored information when power is removed or even momentarily interrupted. Therefore, RAMs are used for temporary storage of data and programs in microcomputers. Since a RAM can be written to or read from rapidly and easily, it is used many times during a program execution for temporary storage, much the same as the accumulator and other registers.

The semiconductor RAM is fabricated from bipolar transistors or MOSFETs (metal-oxide semiconductor field-effect transistors). The most common type is the NMOS (N-channel metal-oxide semiconductor) memory, which can access data in about 100 ns. It is available in two forms: static and dynamic. A static RAM stores data as long as power is applied to the chip. It does this because a static RAM is essentially a flip-flop that will stay either SET or RESET (high or low state) indefinitely provided that power to the circuit is not interrupted. Dynamic RAMs (DRAMs) store data as charges on capacitors. Because these capacitors gradually discharge, the DRAM loses its stored data, making it necessary to recharge the capacitors periodically. The DRAM, therefore, requires additional circuitry to recharge the memory cells. This recharging is referred to as *refreshing the memory*, and it must be done at least every 2 ms. This need for memory refreshing is a great disadvantage. Therefore, the DRAM is economically justifiable only for relatively large memory arrays so that the cost is spread over the large memory.

A new type of RAM, the integrated RAM (iRAM), combines the best features of both static and dynamic RAMs. The iRAM contains the memory cells and the refresh control circuitry on the same chip. This combines the simplicity of static RAMs and the density, cost, and lower power consumption of DRAMs.

## Input/Output Devices

A wide variety of I/O devices, called peripheral devices, may be connected to microcomputers, and they permit the microcomputer to communicate with the outside world. Some are input only, such as the keyboard, joystick, mouse, and other analog-to-digital (AD) converters. Others are output only, such as CRT display monitors, printers, plotters, and other digital-to-analog (DA) devices. Other peripherals are both input and output devices, such as magnetic tape and diskette memory storage.

Input and output devices are similar to memory access. The microprocessor can transfer data to and from peripherals in the same way as it transfers data to and from memory. In fact, external memory is simply another peripheral device. However, when interfacing peripheral devices we encounter some major problems that are not encountered when interfacing memories. The single greatest problem in interfacing I/O devices is that there are so many types; hence there is great diversity in their signals. Not all peripherals naturally conform to I/O standards. They vary tremendously from mechanical to electronic. The signals may be digital or analog, very fast or so slow that changes occur in hours or days. Thus, the interfaces must provide the proper timing and control.

# Interfacing

There are two things to do with a microprocessor—program it and interface it. Before progressing through specific interfacing applications, review the bus concept discussed earlier, as well as timing and control.

Figure 3-6 shows a typical microprocessor with bus arrangements that include addressing capabilities. The address bus provides decoder inputs from the microprocessor. The outputs of the decoders go to the CE lines of the various circuits. Since only one address can appear on the address bus at any given instant, only one external circuit at a time will be enabled.

## Memory Interfacing

The memories are assigned many addresses, because each byte must have its own address. As indicated earlier, the ROM and the RAM are the two general categories of memory devices in

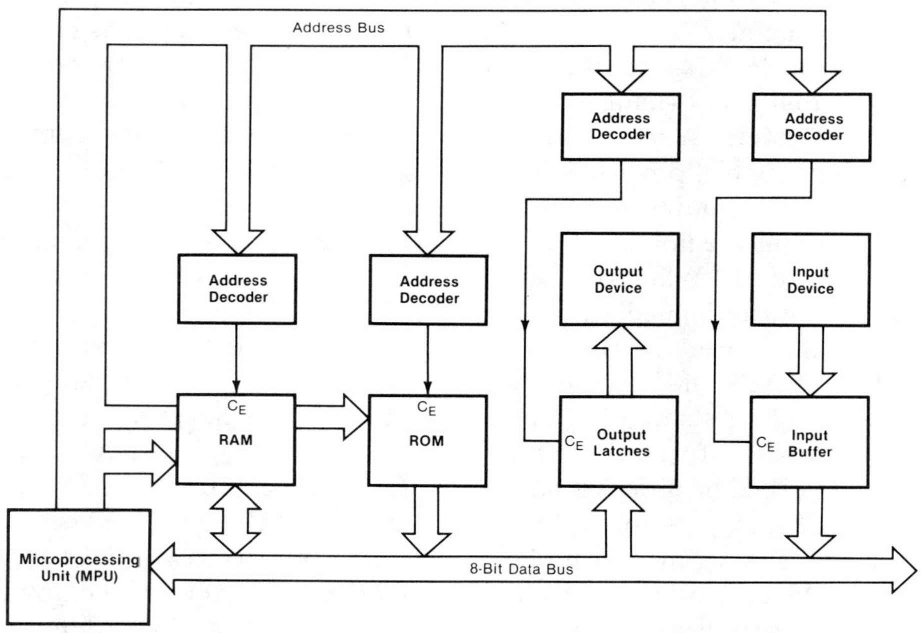


**Figure 3-6.** A typical microprocessor bus arrangement with address capabilities.

common use. To understand interfacing a memory device to a microprocessor, we must read the timing diagrams of the memory device that are on the manufacturer's data sheet. The timing diagrams are shown as a composite waveform of the address connection and not as individual address lines. They are presented in the arrangement shown in Figure 3-7 in order to conserve space and make it easier to understand the waveforms.

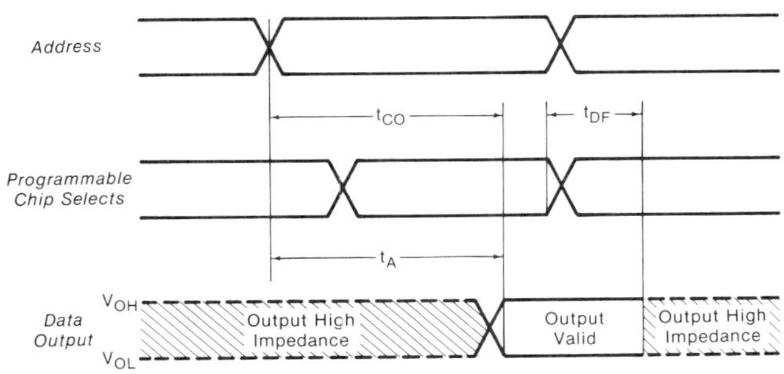


**Figure 3-7.** Typical timing diagram for a ROM.

Notice on the timing diagram that the application of the memory address is the first change that occurs. This then triggers a decoder, as shown in Figure 3-6, which generates the CE input to the memory. The output buffers are then enabled by actuating the OE connection of the memory chip. Then the valid data are output to the data bus. The amount of time that valid data are on the bus varies considerably from one memory device to another. This time required to acquire the information is the access time. The access time for an Intel 2716 EPROM is 450 ns. This time is just for the memory device to access the memory. The decoder must be addressed, and then it must put out a CE signal. The technician performing advanced trouble-shooting should be able to measure these times and compare them with the values on the device's data or specification sheets. These types of measurements are made with an oscilloscope. The techniques are described in Chapter 6.

Very similar to the timing diagram of the EPROM is the one for the static RAM shown in Figure 3-8. The exception is that the WE (write enable) replaces the OE. In some microprocessors, the WE pin is labeled R/W and has exactly the same function.

The READ cycle is almost identical to that of the EPROM, though the access time is much shorter. The WRITE timing diagram shows that again the first thing that happens is that the memory address is sent out and decoded to enable the device. The WRITE signal (WE low) then follows and indicates that the data to be written into the memory are available on the data bus. The memory READ cycle time and

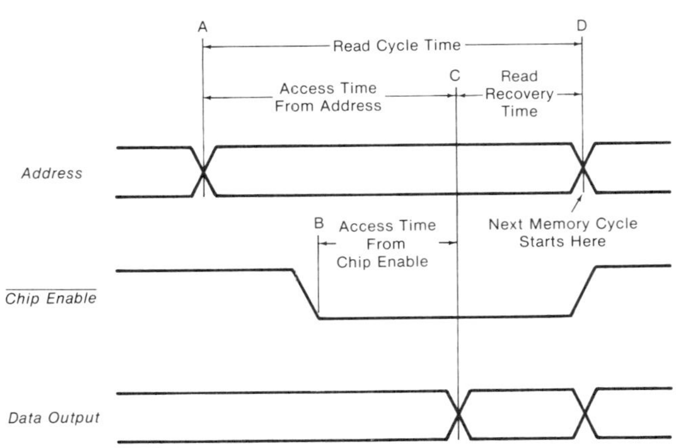


(A) READ cycle.

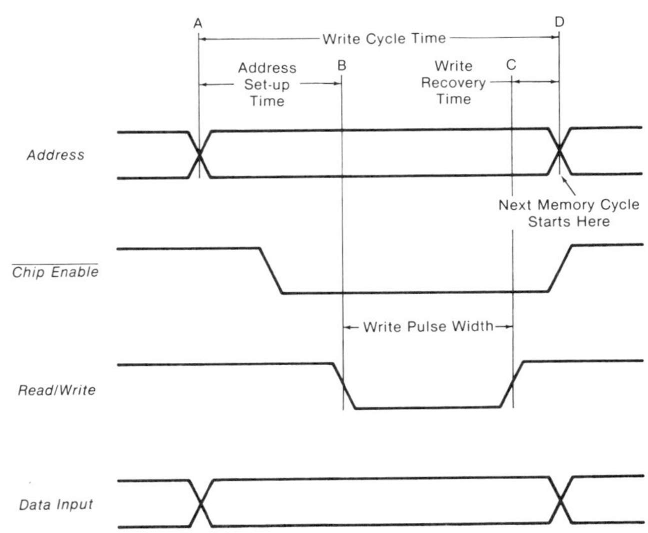


(B) WRITE cycle.

**Figure 3-8.** Typical timing diagram for RAM.

the WRITE cycle time may be different for various memory chips.

The address decoder in Figure 3-6 is a device that is, or contains, a digital combinational logic circuit. It indicates that a particular area of memory is being addressed by the microprocessor, which in turn activates one or more outputs for that area of memory. A very popular decoder for a wide variety of microprocessor-based equipment is the 74LS138, or Intel 8205, which is a 3-to-8-line decoder. The outputs of this device are active low. Also, only one output becomes active when it is enabled and the input for that particular output is addressed.

## Basic Input/Output Concepts

Input/output interfacing is an important part of microprocessor-based equipment operation. A basic understanding of these concepts is essential for success in troubleshooting this type of equipment. The point at which an I/O device delivers data to or accepts data from the microprocessor is called the *interface point*. At this point, the device must either be or become TTL (transistor-transistor logic) compatible.

### Input Device Interfacing

Most input devices are actually some form of switch—a simple toggle, pushbutton, a rotary from a reciprocating machine, or an electronic switch such as an AD converter. Switch bounce has been a problem in the past with hard-wired TTL logic circuits, and the switches have had to be double-throw types with NAND gate or inverters to form a debounce circuit. Today, microprocessor-based inputs require only a single-pole single-throw (SPST) switch with a pull-up resistor to +5 V to make it TTL compatible. Any contact bounce is taken care of with software in the microprocessor. This signal is then input to the data bus with a digital switch, such as the 74LS244 octal buffer. The decoder selects this particular device, then enables it by applying a logic 0 to its instrobe, which turns on (enables) the three-state buffers. This strobe lasts for about 500 ns. Once enabled, these buffers connect data to the microprocessor's data bus from the input switch or device.

### Output Device Interfacing

The simple output devices are usually display devices and coils or relays or solenoids. The most common of the display devices is the light-emitting diode (LED). Most LEDs require 1.65 V at 10 mA for full brilliance. To operate them from the TTL 5-V supply requires a 330-Ω dropping resistor.

A 7-segment LED display has an internal structure that is merely a single LED for each segment. The LED's requirement for relatively high current is a disadvantage because common TTL logic gate drivers are not sufficient. High-current drivers or Darlington pairs are used instead. A number of ICs are especially designed to drive the 7-segment display. A typical example is the 7447 shown in Figure 3-9. It is a binary-coded-decimal (BCD)-to-7-segment-display decoder/driver. The 7447 decoder/driver receives a 4-bit binary number at inputs A, B, C, and D and provides proper output combinations at a through g to form the numerals 0 through 9 and the six characters. The 7447 does not have latch capability to store the data and keep the segments on. Thus a separate latch must be used to hold the data. A disadvantage of using the 7447 with a microprocessor system is that it is limited to the 16 symbol combinations provided by the 7447 decoder/driver instead of the 256 possible combinations available on the 7-segment display plus the decimal point (Dp).

A more versatile way of driving a low-current 7-segment LED is to drive the display directly by an 8-bit bistable latch as shown in Figure 3-10. The seven segments and decimal point can now be controlled directly by the data bus of the microprocessor.

The LCD is common where the output needs high visibility in bright light and low power requirements. LCDs do not generate light; they

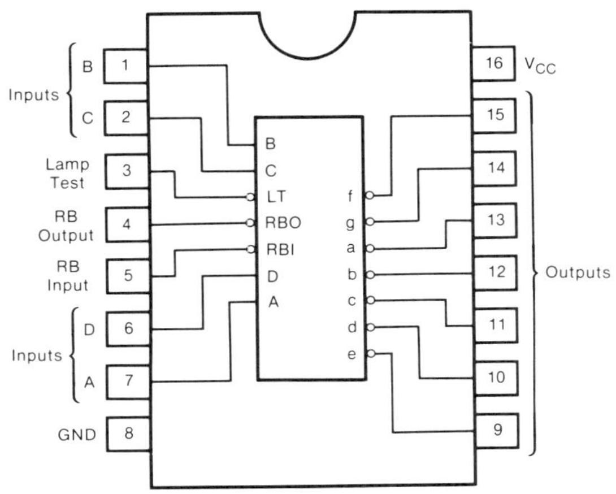


(A) The 7447 7-segment decoder/driver.

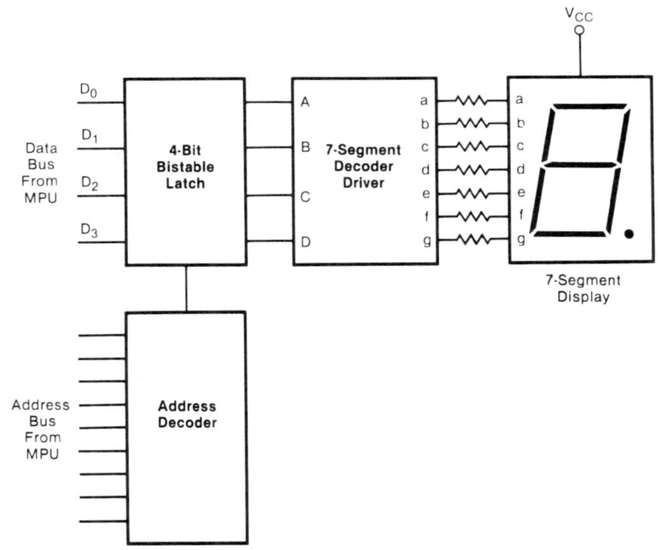


(B) Using the decoder/driver.

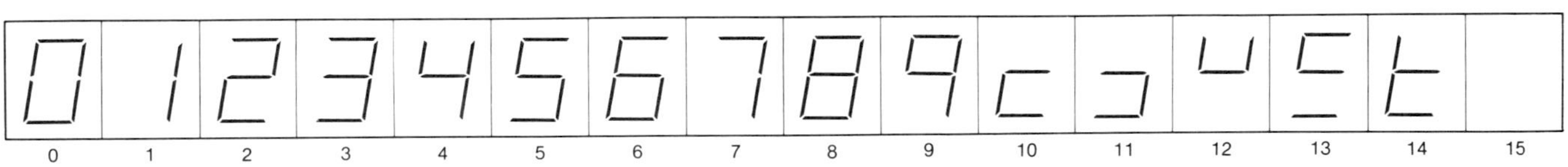


(C) The 7447 7-segment decoder/driver resultant display.

**Figure 3-9.** 7-segment LED and 7447 driver.

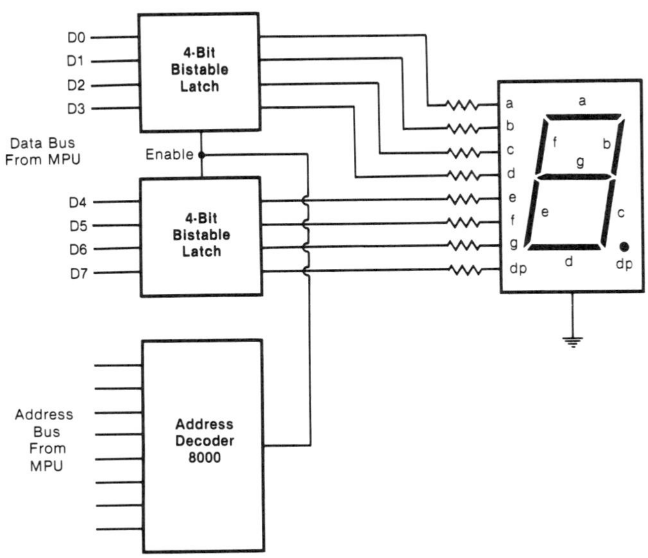


(A) A more versatile way of driving a 7-segment display.

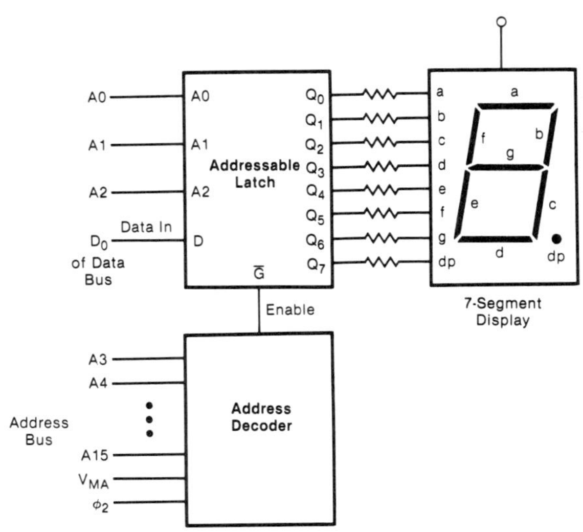


(B) Using the addressable latch to drive a 7-segment display.

**Figure 3-10.** 7-segment LED driven directly by MPU.

reflect or transmit light from an external source. The LCD cells are arranged in the same 7-segment format as the LED for numerical display. There are two types of LCD display:

1. The reflective type has a mirrored background and appears to light up when a segment is aligned.
2. The absorption type has a black background, so it appears to darken when a segment is aligned.

The segments require an AC excitation square wave between 30 Hz and a few kilohertz at approximately TTL logic levels.

If the simple output is not a display device, then it will more than likely be the coil of a relay or solenoid. If the microprocessor is used to control power to an AC load operating from the 60-Hz power line, it is necessary to isolate the logic from the load circuit that it is controlling. This necessary isolation can be provided with either an electromagnetic relay or a solid-state relay. The electromagnetic relay is available in "reed" form in dual in-line packages just like other digital ICs. The solid-state relay is usually a form of optoisolator or optocoupler. In most cases, the logic signal first goes to a buffer, such as the 7406 hex buffer. It has an open-collector transistor output that can drive (sink) up to 40 mA of current when on and withstand up to 30 V when off. For higher currents, larger buffers are available. However, when these larger buffers are used to control current to an inductive load, such as a relay, solenoid, or motor, they have to be protected by a damper diode against the inductive kick that high voltages produce. Figure 3-11 shows a 7406 driving a relay.

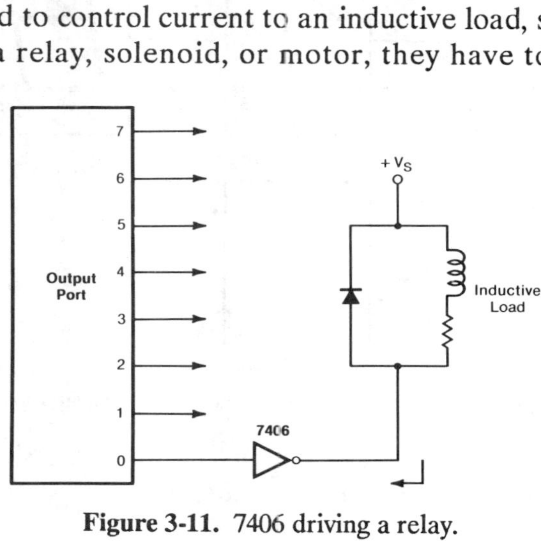


**Figure 3-11.** 7406 driving a relay.

There is a class of ICs that bridges the gap between the disparate worlds of electrical power and digital electronics. These "smart-power" devices use low-voltage signals in the logic section to control high voltages and currents. This need to carry high voltages as well as high currents is the big challenge for smart-power technology. In addition to reducing an electrical system's bulk and cost, these chips will inherently replace unreliable electromechanical relays. Instead of controlling the high power at the interface point, they send a digital signal through a small communications wire loop that is connected to a smart-power device at each load.

## Systems of Input/Output Interfacing

Of particular importance in I/O interface systems are the methods used to control the transfer of information to and from the microprocessor to memory or peripheral devices. Three methods are available to control the transfer:

1. Memory-mapped I/O
2. I/O mapped control, sometimes referred to as isolated I/O or direct I/O
3. Attached I/O

*Memory-mapped* I/O treats the I/O device as if it were a location in memory. A portion of memory must first be set aside for the I/O devices. These memory addresses are not to be used by the normal memory function. A simple method of implementing memory-mapped I/O control is shown in Figure 3-12.

The I/O port decoder was mentioned when memory interfacing was discussed earlier. An eight-input NAND gate is usually used to decode the address within the range and develop the page selection strobe (PAGE CF). This strobe is used to enable the port decoder at $E_1$ and allow the selection of one of eight different devices with inputs A, B, and C. Memory locations DF00H through DFFFH are designated for the

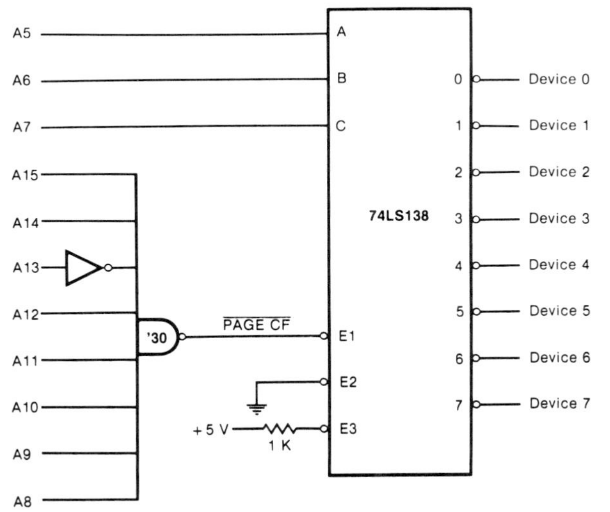


(A) The memory-mapped I/O port decoder.

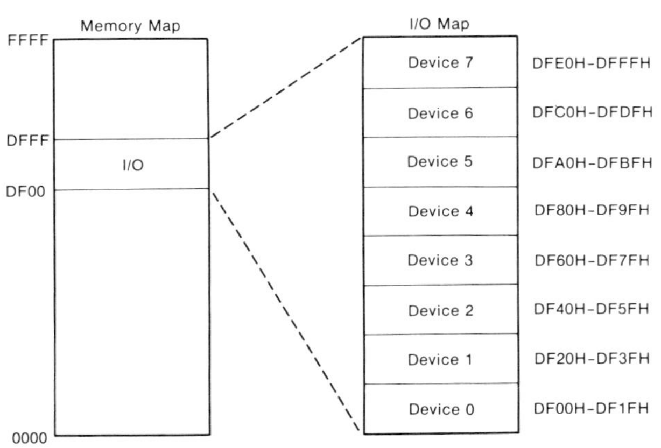


(B) An example memory and I/O map for memory-mapped I/O page DFXX.

**Figure 3-12.** A memory-mapped I/O port decoder and the memory map for it.

I/O port of memory (that is, one page of memory). This may appear to be an excessive amount of memory to dedicate to the use of eight I/O devices. However, when weighed against the cost of decoding more bits of memory, the use of more memory addresses than needed is less expensive, provided that there is ample overall memory. If memory locations are at a premium, then more address decoding is required. Device 0 is active for any of the addresses in the DF00H to DF1FH range, device 1 for addresses DF20H to DF3FH, and so on through device 7, with each device overlaying 32 memory locations. The program can activate a particular device by addressing any location in its memory range.

Memory-mapped I/Os can be used with any microprocessor. However, some microprocessors, such as the 8086 family, allow you to set up separate address spaces for input ports and for output ports. A microprocessor system having separate address spaces for input and output ports is called *direct I/O.*

With direct I/O mapped control, the input and output ports are accessed directly with the IN and OUT instructions. The I/O is given priority to use the data bus over memory. An I/O request control signal is brought into the I/O unit from the CPU, which indicates that the I/O unit is activated. The I/O unit, port, or even the particular I/O bit is then selected by using any address on the memory address bus. The memory itself is kept inactive during the time the I/O request signal is active. The 8086 IN and OUT instructions each have two forms: fixed port and variable port. For fixed-port instructions, an 8-bit port address is written as part of the instructions. The instruction IN AL, 26H would copy a byte of data from port 26H to the AL register. To use the variable-port input or output operation, first load the 16-bit port address into the DX register. Then use the instruction IN AL, DX to copy a byte from the addressed port to the AL register. When the IN or OUT instructions are used on the 8086 to access a port, the port address is set out directly from the microprocessor on address lines AD0–D15, and 0s are sent out on lines A16A19. An 8086 using direct I/O is designed so that the separate address space for I/O ports is enabled when the $M/\overline{IO}$ signal from the microprocessor is low and the RAM and ROM decoders are enabled when the $M/\overline{IO}$ is high. A memory cycle is therefore in progress when the $M/\overline{IO}$ is high.

In general, memory-mapped I/O is best suited for systems that use complex interface chips, whereas direct I/O is better for systems that use small-scale or medium-scale ICs.

The third method, *attached I/O,* is a useful alternative for small systems where the number of chips must be kept to a minimum. In this method, the I/O ports and control circuitry are part of the CPU or memory source. Typical examples of applications for attached I/O are the I/O controller for consumer appliances and tools, and video games where speed is not essential but simplicity is.

## Processing Interrupts

An *interrupt* is an extremely important microprocessor hardware concept that allows an external device to get the attention of the microprocessor. Most of the time, the microprocessor is executing the main program routine. Many interrupts are an input signal to the CPU that can directly alter the sequence of operations at the hexadecimal level. The interrupts tells the currently executing program to branch or jump to a subroutine. Usually, these are pin connections on the microprocessor that when activated cause the interrupt to occur. This is referred to as a *hexadecimal interrupt.*

A keyboard interface is a common application of an interrupt. Without an interrupt system, the microprocessor would spend most of its time scanning the keyboard looking for the operator to strike a key. With an interrupt system, the microprocessor only spends time with the keyboard when a key is actually pressed. Then this interrupt calls the subprogram—called the *interrupt service routine*, which is the software that responds to the interrupt—from the hexadecimal. In this example, the interrupt service routine determines exactly which pressed key has caused the interrupt.

Recall the two basic interrupt types: NMI and INT. Because the NMI input cannot be disabled, it handles the highest-priority interrupt conditions, such as a power failure. If a power line failure is detected, the power supply usually has several milliseconds remaining before the voltage falls to a level that will result in loss of data in the RAM and microprocessor registers. This is ample time for the register information to be stored in memory and for a battery backup to be switched on.

The maskable interrupt is more useful than the nonmaskable because it can be program controlled by incorporating a mask in the program, which allows it to be recognized or not recognized by the system. After an interrupt request is received, the software can also provide a polling routine that polls the I/O devices from the highest to the lowest priority. When the device that has requested an interrupt has been found by the polling routine, the program jumps to the interrupt service routine and data are transferred to or from the device. After the transfer is complete, the program will return to poll the remaining devices.

A second source of an interrupt on the 8086 microprocessor is the execution of the interrupt instruction INT. This is called a *software interrupt.* The 8086 can also be interrupted as the result of an instruction in the program that tells the microprocessor to do something that it cannot do, such as divide a number by 0. This is a software interrupt also, but it is referred to as a *conditional interrupt*. At the end of each instruction cycle, the 8086 microprocessor checks to see if any interrupts have been requested. Many common microprocessors have only hardware interrupt request ability available.

## Direct Memory Access

The only data transfer that we have considered up to now is transfer between the microprocessor and the I/O devices, especially memory, and between registers within the microprocessor. This method is called *programmed transfer*; that is, the microprocessor executes a sequence of instructions for transferring the data. In this method, the data do not pass from memory to a peripheral device, or vice versa, because they must first pass to and through the microprocessor. Often time is taken up by the microprocessor fetching instructions as the data move from register to register through the microprocessor.

Direct memory access offers a substantial increase in speed over programmed transfer in moving data to or from RAM and a peripheral device. Here, the data are transferred directly between the peripheral device and the memory. DMA uses a dedicated hardware device called a DMA controller, which temporarily takes control of the address bus, the data bus, and the control bus from the microprocessor. The DMA controller then transfers data bytes from (to) the peripheral device to (from) memory totally at hardware speed. The DMA controller is essentially a specialized processor that takes over memory and I/O space. It is almost as complex as a microprocessor; that is, it obtains an address register to select the desired transfer location and control logic that provide both memory and I/O devices with appropriate control signals. The DMA techniques are best suited for data transfer from memory to high-speed devices, such as magnetic disk and tapes, high-speed printers, and CRT terminals.

# Comparing Microprocessors

In comparing microprocessors, it is unfair for us to state that one microprocessor is better than another. Many factors—speed, cost, power requirement, instruction set, internal registers, memory address range, bits required for a particular task, and the control bus functions—must be considered. Usually, a specific microprocessor is designed for a specific application or segment of the market. Therefore, hardware and software aspects also need to be analyzed.

## Four-Bit Microprocessors

The TMS1000 (Texas Instruments) is the most popular 4-bit microprocessor, though in reality it is a self-contained microcomputer. It has a self-contained memory and is thus limited to using what memory is available. Because the TMS1000 is widely used in games, microwave ovens, and washing machines, its price has dropped quite low.

Another popular 4-bit microprocessor is the 4040 (Intel), which is an updated version of the original 4004 (Intel) conceived in 1971 as a calculator controller.

The word size of the 4-bit microprocessor proved to be too restrictive in many applications, especially those that handle alphabetic information. It normally takes 6 or 7 bits to represent 26 letters plus several special characters. Thus, the 8008 8-bit microprocessor was introduced by Intel.

## Eight-Bit Microprocessors

The first 8-bit microprocessor, the Intel 8008, had two drawbacks: it was slow and required a negative power supply that was not TTL compatible. Both of these limitations could be blamed on the P-channel metal-oxide semiconductor (PMOS) technology.

About 1972, there was a breakthrough in fabrication technology and NMOS logic circuitry became feasible. This circuitry increased the speed by about 25 times and allowed the use of a positive power supply, which thus led to the birth of the Intel 8080 in 1973, and the microprocessor age had truly begun. In comparing other 8-bit microprocessors, we find that their internal architectures differ slightly, but their functions and applications are quite similar. Virtually all 8-bit microprocessors have a 16-bit address bus, and all have an 8-bit data bus. Most of them operate at a clock speed from 1 to 5 MHz.

Now, let's look at some popular microprocessors, beginning with Intel's 8080 and 8085. Refer to Figure 3-13 for the pinout of each microprocessor.

### Intel 8080

In addition to the 8080, two other chips are necessary for completing the CPU. One is the 8224 clock generator chip, and the other is the 8228 system controller, which may be optional in some applications. The 8224 clock generator supplies the two-phase, nonoverlapping clock signal that the 8080 needs. It also contains the necessary logic circuitry for decoding the status strobe signal. The 8228 system controller is used to decode the control signals that are multiplexed on the data bus and to act as a bidirectional

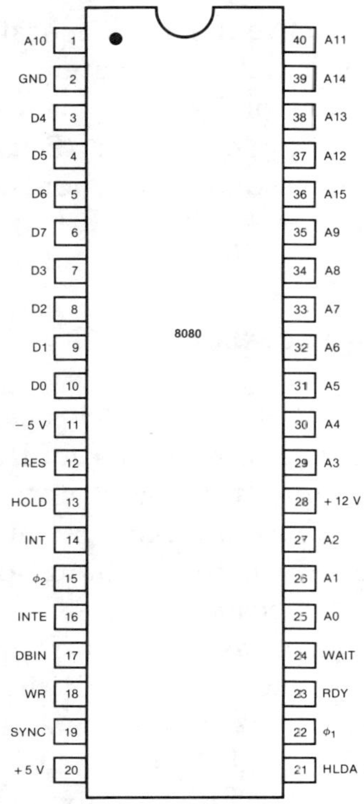


(A) 8080.

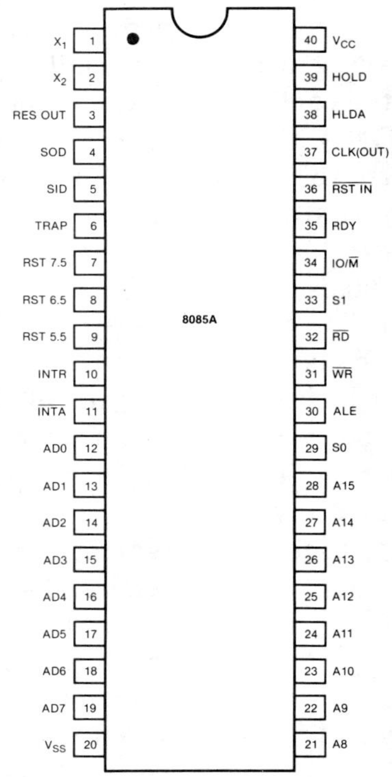


(B) 8085A.

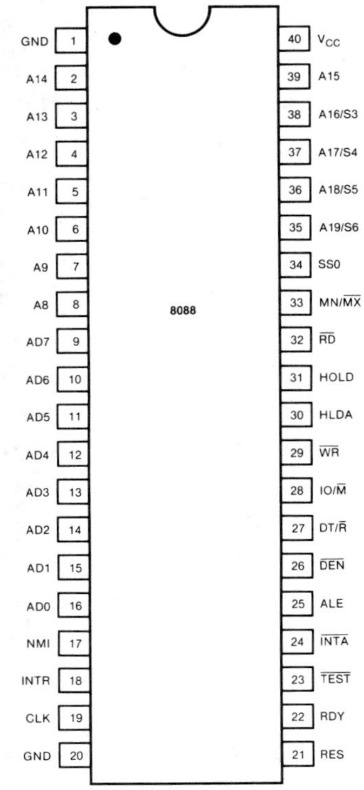


(C) 8088.

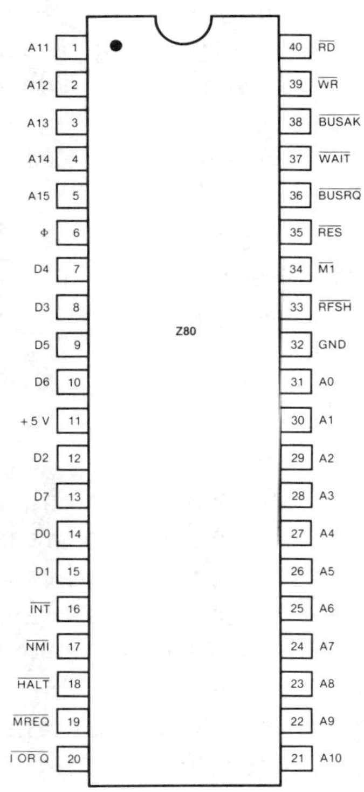


(D) Z-80.

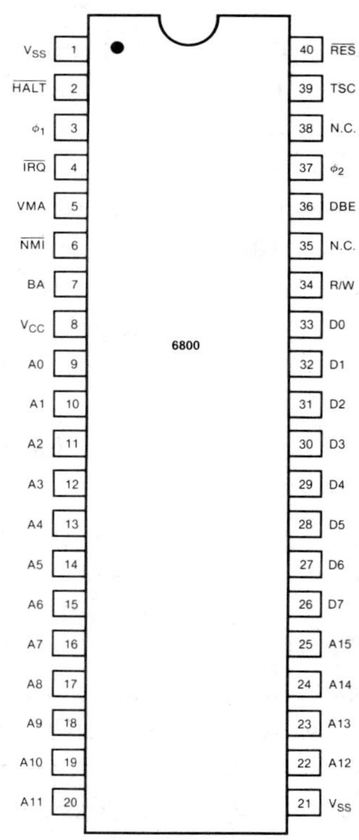


(E) 6800.

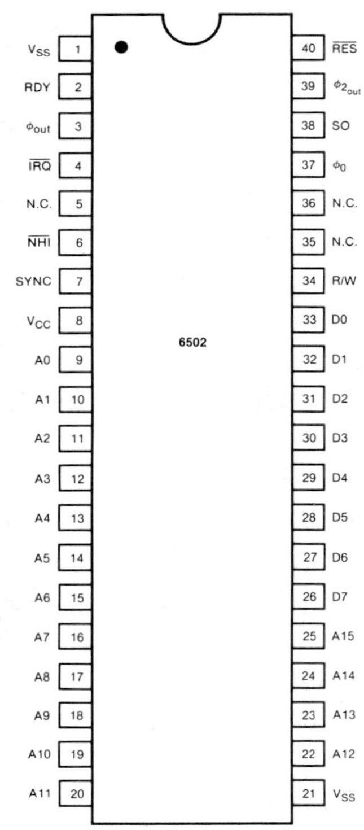


(F) 6502.

**Figure 3-13.** Pin assignments for 8-bit microprocessor.

buffer to boost signal power of the data bus signals.

The 8080 has six 8-bit general-purpose registers; an accumulator and an instruction register, each of which is 8 bits; and a 16-bit program counter and stack pointer. It has a program status word with five flags: carry, zero, sign, parity, and decimal adjust. The 8080 requires a power supply with three voltages: +5, -5, and +12 V.

### Intel 8085A

The 8085A is an 8-bit general-purpose microprocessor suitable for many applications. It needs a latch to catch the multiplexed low address lines (on the eight address/data lines) when the address latch enable (ALE) signal is active. No elaborate clock generator chip is necessary. It has three pins that control or present the clock signal. Pins 1 and 2 ($X_1$ and $X_2$) are used to connect a crystal to control the clock frequency, or a square wave can be input on pin 1 while leaving pin 2 floating. The clock output, CLK(OUT), is available on pin 37.

The 8085A contains 10 principal registers: a set of six 8-bit general-purpose registers that can be used in pairs as three 16-bit registers, which the programmer can utilize in any manner; and two 8-bit and two 16-bit special-purpose registers used for accumulating the results from arithmetic and logic instructions and for housekeeping (that is, tasks that are required but normally occur without the intervention of the programmer). Accumulator A in the 8085A, like most other 8-bit microprocessors, is used to accumulate the answer after almost every arithmetic and logic operation. The flag register contains five bits: the sign flat bit, the zero flag bit, the auxiliary carry flag bit, the parity flag bit, and the carry flag bit. The PC, SP, and the increment/decrement address latch are all 16-bit. This leaves the buffers on the address bus and the data/address bus, each of which are 8-bit. The power supply requirements are a well-regulated 5 V at approximately 170 mA of circuit. Pin 40 is +5 V, and pin 20 is ground (0 V).

### Intel 8088

The 8088 is an 8-bit version of the popular 16-bit 8086. It has fewer data lines, but retains all of the processing features of the 8086 and is fully software compatible; that is, it has the same instruction set. Like the 8080 and 8085A, the 8080 has eight data lines, although it can manipulate 16-bit operands by handling them in two 8-bit bytes. It has a 20-bit address bus, so it can address any one of 1,048,576 bytes in memory. The 8088 requires an external clock generator or an external TTL compatible clock for its operation. An 8284A divides the crystal frequency by 3 for the clock input to the microprocessor. Thus, a 15-MHz crystal is attached to it, which provides the typical 5.0-MHz clock frequency for the 8088.

### Zilog Z-80

The Z-80 microprocessor was developed as an enhancement to the 8080. It has similar architecture to the 8080 and the 8085 and can execute machine language programs written for either one. It requires an externally generated single-phase clock signal input up to 2.5 MHz, with the Z-80A able to run at 4 MHz.

There are 22 principal registers: two sets of 8-bit registers, each set containing six general-purpose registers; an accumulator; and a status register. The two special 8-bit registers are the interrupt vector (IV) and the memory refresh counter (MRC). The MRC is used when DRAM is used as part of the memory. The 16-bit registers are the PC, the two index registers (IRX and IRY), and the SP. The Z-80A can perform essentially the same operations as the 8085 plus many additional operations at a somewhat higher speed, because it has one of the largest instruction sets among the 8-bit microprocessors. It has a 16-bit address bus and a 13-line control bus. The Z-80A requires a +5-V supply.

### Motorola 6800

The 6800 microprocessor was introduced in 1974. Though its architecture is simple, it has been widely used in industrial applications. The 6800 has a 16-bit address bus, an 8-bit data bus, and an 8-bit control bus. It requires a two-phase externally generated clock signal and a +5-V supply.

The registers for the 6800 include two accumulators (A and B) and a status register, all 8-bit. It has three 16-bit registers, including the PC, index register, and SP.

The 6800 has a complete family of LSI support chips used to interface the 6800 to a variety of I/O devices. The ease of interfacing and simplicity are the reasons for the 6800's popularity.

### MOS Technology 6502

The 6502 was originally designed to have a complete two-phase crystal oscillator inside the chip. However, that proved to be unreliable, so later versions have an external clock. It has six principal registers, which include the common 8-bits, accumulator, two index registers, status register, and SP. An 8-bit SP is sufficient to determine the address of the stack because the 6502 uses page 01 of memory for the stack. It also has a 16-bit PC. The basic 6502 operates at frequencies up to 1 MHz. However, there are faster versions available; the 6502A and 6502B operate at 2 MHz and 3 MHz, respectively.

The ALU of the 6502 has the same basic operations as the other 8-bit microprocessors. It does, however, have an advantage in the execution of BCD arithmetic because it does not require a decimal-adjust to correct BCD results if the D flag has been set. Like the Z-80 and 6800, the 6502 microprocessor requires a single 5-V power supply.

## Sixteen-Bit Microprocessors

From the original 4-bit 4004 microprocessor, IC technology development has moved toward the microprocessor handling larger word sizes. With 16-bit words, more information is carried and larger numbers can be more easily represented. The term *16-bit* means that its ALU, internal registers, and most of its instructions are designed to work with 16-bit binary words. There is more variation in the architecture of 16-bit microprocessors. The greatest variation is that the width of the data bus may be 8 bits or 16 bits. Since much of the available peripheral equipment, both memories and I/O devices, is 8-bit data bus, we can save much time and money by using existing designs. Also, all of the 16-bit microprocessors currently in use incorporate additional features, such as instructions for multiplication, division, and block data manipulation. This reduces the software coding time. The most prominent 16-bit microprocessors are the Intel 8086, the Zilog Z-8000, and the Motorola 68000. Their pinouts are given in Figure 3-14.

### Intel 8086

The 8086 is an extension of the 8085, though the 8088 is its contemporary 8-bit version. The 8086 has a 16-bit data bus, so it can read data from or write data to memory and I/O ports, either 16 bits or 8 bits at a time. It has a 20-bit address bus and can thus address any one of 1,048,576 (8-bit byte) memory locations by storing 16-bit words in two consecutive memory locations. Whether the entire word is read in one operation or two depends on whether the first byte is at an even or an odd location. If the first byte of a 16-bit word is at an even address, it will be read in one operation; if at an odd address, its first byte will be read in one operation and the second in another operation.

The 8086 CPU is divided into two independently functional parts: the BIU and the EU. The BIU sends out addresses, fetches instructions from memory, and reads data from and writes data to ports and memory. That is, it handles all transfers of data and addresses on the buses for the EU. The EU tells the BIU where to fetch instructions or data, decodes instructions, and executes instructions.

The 8086 has eight general-purpose registers, which may be used individually for 8-bit temporary storage or in certain pairs to store 16-bit words. The SP register is 16 bits, which the 8086 allows to use a 64-kbyte section of memory

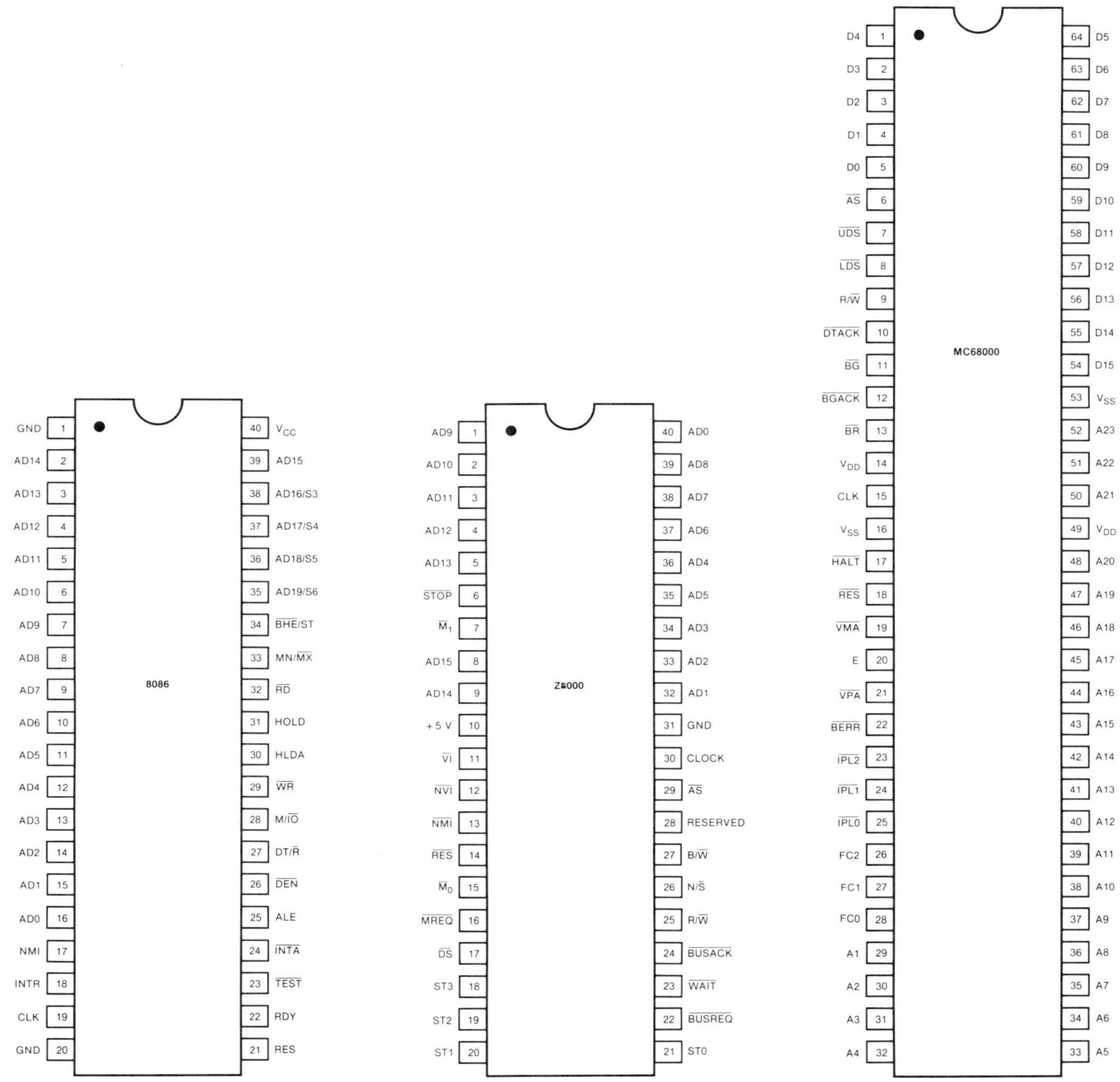


(A) 8086. (B) Z8000. (C) MC68000.

**Figure 3-14.** Pin assignments of three 16-bit microprocessors.

for the stack. The EU contains a 6-bit flag register.

Another feature of the 8086 is that it is able to perform arithmetic operations directly on numbers that are in ASCII-coded form, which significantly reduces processing time.

## Zilog Z-8000

The Z-8000 is available in two packages: the 8001 and the 8002. The 8001 is a 48-pin package and is the maximum system chip; it can control more memory and has greater capabilities than the

8002. The 8002 is a simplified 40-pin version that multiplexes its data bus and address bus on the same 16 pins by making use of the control signals address strobe (AS) and data strobe (DS). A group of 4 bits called status (ST) signals is used to encode and decode any of 16 conditions, such as interrupt granted and memory or I/O strobe signals. It has a single-phase clock and requires a single 5-V power supply. The Z-8000 has 16 general-purpose registers that can be used in combination as 8-bit, 16-bit, 32-bit, or 64-bit registers. These registers can perform arithmetic operations on 32-bit numbers using a single instruction.

### Motorola 68000

The 68000 is a 16-bit microprocessor with a 24-bit address bus and a 16-bit data bus. However, unlike the Intel and Zilog microprocessors, these two buses are not multiplexed on the same pins. This results in a 64-pin package. With its 24-bit PC, it can address 16 Mbytes of memory. Internally, the 68000 microprocessor has seventeen 32-bit, multifunction registers that can be used for 8-bit, 16-bit, and 32-bit operations. Each one can also be used as an index register.

Variation in 16-bit architecture is greater than in the standard 8-bit microprocessor. However, an understanding of the 8-bit machine makes any of the 16-bit machines easier to group. The 16-bit microprocessors are not that different, just more complicated. Troubleshooting them is quite similar to troubleshooting 8-bit microprocessors.

## Bit-Slice Microprocessors

The fastest logic families are still bipolar devices, in particular, low-power Schottky TTL and ECL. Their disadvantage is high power dissipation, which prevents them from being implemented into a single-chip microprocessor design. The bit-slice technique of microprocessor construction consists of using a 4-bit ALU, registers, buses, flags, and multiplexers to form a basic bit-slice device. This device is then connected in parallel with other bit-slice devices to handle larger word sizes. This technology is appropriate only when an application requires extremely high speed, very wide word widths, or a tailored instruction set. To perform a useful operation equivalent to a single machine language instruction, the bit-slice device must go through microinstruction. A special device, called a *microprogrammer sequencer* is in control of sequencing through these microinstructions in the ROM. The bit-slice microprocessor may contain tens of thousands of these microinstructions.

## Single-Component Microcomputers

When a limited amount of memory is required for consumer and industrial electronics applications and the chip count must be kept low, then single-chip computers are attractive to the designer. These devices typically contain an 8-bit microprocessor, ROM or EPROM for the program storage, RAM for the data storage, and several I/O ports. The clock is usually internal as well. Some examples of these single-chip computers are the 4-bit TMS1000 from Texas Instruments and the Intel 8048, which contains an 8-bit microprocessor, 1K of unalterable ROM, 64 bytes of RAM, and three 8-bit ports. Intel also makes a single-chip 16-bit microcomputer, the 8096, which is finding wide use in certain types of electronic instruments.

Often, small systems can be designed with one of these microcomputers and very few other components, which decreases expense, development time, and problems for the troubleshooter in maintaining it.

An understanding of the microprocessor and microcomputer is essential for repairing digital electronic systems that contain these devices. Other servicing skills are also necessary to troubleshoot microprocessor circuits. In Chapter 4, we will look at the main LSI chips found in a microprocessor-controlled circuit with the purpose of developing the needed troubleshooting skills.

# 4

# Meet the Main Chips

If you examine a system that uses a microprocessor for control of operation, you will encounter four or five large-scale integrated circuits (LSIs). These circuits are usually 40-pin chips, although 24-pin and 64-pin chips are occasionally used. They will invariably be the central processing unit (CPU), peripheral interface adapter (PIA), video display generator (VDG), synchronous address multiplexer (SAM), and the asynchronous communications interface adapter (ACIA). These chips plus ROM, RAM, and a power supply make up not only a personal computer but most of the microprocessor-based industrial control systems, as well as the computer-controlled car (CCC). Various companies use different names for their interface devices, though they may perform essentially the same function. Many of the servicing skills used to repair general electronic circuits are also used to troubleshoot microprocessor-based systems. In this chapter, we will look at the operation of the main LSI chips, with troubleshooting in mind.

## Central Processing Unit

As we saw in Chapter 3, the arithmetic logic unit (ALU) and the control unit are usually combined to form the CPU. The CPU is truly the heart of the computer system because it is connected to all of the other chips with lines, like arteries and veins, called *buses*. The various components in the CPU are connected together with lines called *internal buses*. The lines connecting the CPU to the remainder of the microprocessor system chips are called *external buses*. A CPU includes the microprocessor and any additional components that it requires, such as a clock chip or bus buffers, controller, and arbiter.

The CPU does not have many functions, although it does control the operation of the system. Without it, there is no computer system. It fetches binary-coded instructions from memory, decodes the instructions into a series of simple actions, and carries out these actions. The CPU contains an ALU, the word length of which determines the microprocessor's word length classification (that is, 4-bit, 8-bit, or 16-bit). The ALU can perform simple actions such as add, subtract, logical OR, AND, invert, or Exclusive-OR operations on binary words as it is instructed. A program or address counter is used to hold the address of the next instruction of data to be fetched from memory. Unless otherwise commanded, the CPU sequences through the set of instructions of the program. The ALU, counters, and registers are connected together by the three buses: the address bus, the data bus, and the control bus.

At the end of Chapter 3, we looked at the pinout diagrams of several popular microprocessor CPUs. Here we will look at two of these in detail: the Motorola 6800 as an example of an 8-bit microprocessor, and the 8086 as an example of a 16-bit microprocessor.

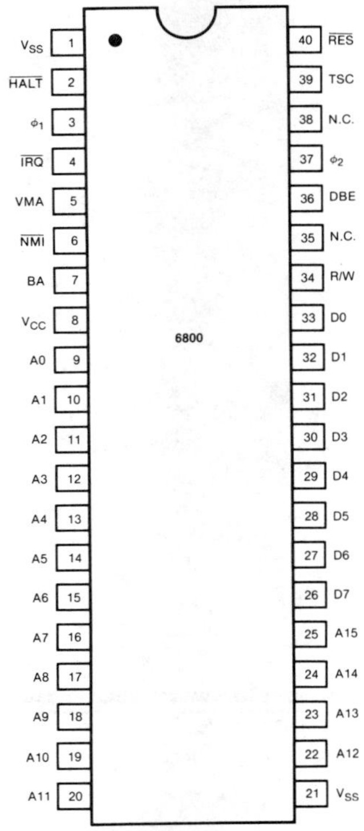


**Figure 4-1.** 6800 pinout diagram.

# The Basic 6800 Microprocessor

The pinout diagram of the Motorola 6800 is shown again in Figure 4-1; its internal architecture is shown in Figure 4-2.

A basic 6800 system may consist of the MC6800 microprocessor, ROM (MC6830), RAM (MC6810), ACIA (MC6850), and a PIA (MC6820). To know the way each of these units works together, we must know the hardware features of each chip. For a component to be a member of a particular microprocessor family and to ensure compatibility, it must meet specific system standards—in this case, the MC6800 family. These standards make it convenient for the other main LSI chips to interface, or communicate, with the microprocessor chip directly. The system components are interconnected by an 8-line data bus, a 16-line address bus, and a 9-wire control bus.

## Three-State Bus Concepts

As do most microprocessors, the 6800 system uses the three-state bus technique to communicate with other components. It depends on the ability of 11 of the interconnected components to be switched to a third state, which presents a very high impedance to the bus. This switching allows only one component at a time to be selected to drive the bus. The selected component causes a high or low to be placed on each line of the bus, and all unselected components must place their bus drivers in the high-impedance state. Also, one of the other components on the bus is enabled to read the data. The microprocesor's read/write (R/W) and valid memory address (VMA) lines control this three-state switching along with the address bus. All of the 6800 family components have this three-state capability in their data output circuits.

The MC6880 is a three-state bus driver/receiver that can interface external devices that are not a part of the 6800 family to the bus. It is available in dual in-line (DIP) packages and is a quad integrated circuit (IC) with four driver inputs to the bus, four receiver outputs from the bus, and four connections to the bus. Its pin connections and internal structure are shown in Figure 4-3. The 6880 may be used when the system is larger and more than 10 MOS (metal-

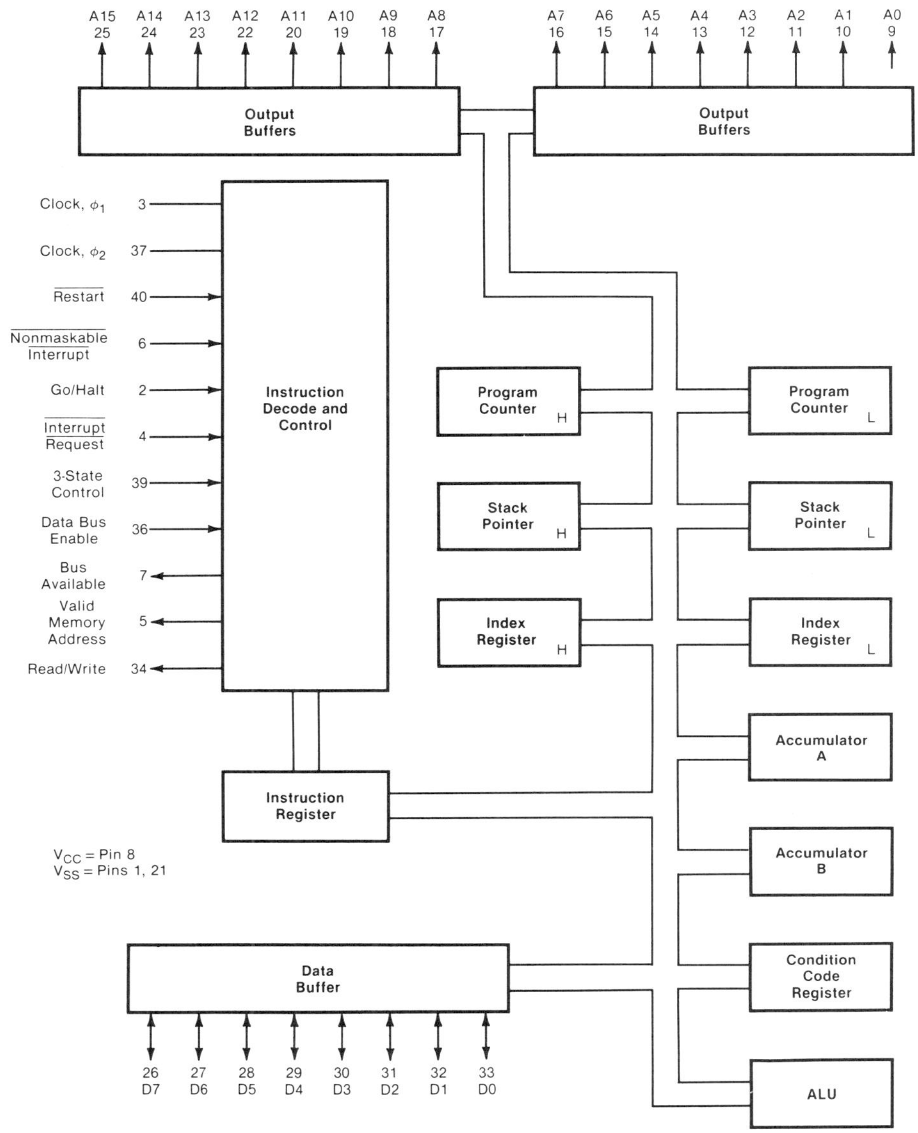


**Figure 4-2.** Block diagram of 6800 microprocessor.

oxide semiconductor) or 1 TTL (transistor-transistor logic) load is used. You can determine the directions of transmission by the R/W line from the microprocessor by tying pins 1 and 15 (receiver enable input and driver enable input) together and by making pairs of pins 12 and 14, 9 and 11, 7 and 5, and 4 and 2 (driver inputs and receiver outputs). For example, if you use this configuration with RAM, the driver enable is enabled when the microprocessor is reading memory (R/W = 1). The receiver enable input is enabled when the microprocessor is writing to memory (R/W = 0).

The outputs of these three-state devices can be tested with a VOM/DVM, logic probe, or an oscilloscope (see Chapter 6) to see if it is either in a steady state (vaild 1 or 0) or high-impedance (floating) state. The high-impedance state occurs when the device is in a disabled state. The disabling may be intentional, due to the disabling signal, or the result of a defect in the chip. Learning to analyze the valid highs and lows (discussed in Chapter 1) and floating status of signals is one of the most important trouble-shooting techniques in digital microprocessor systems.

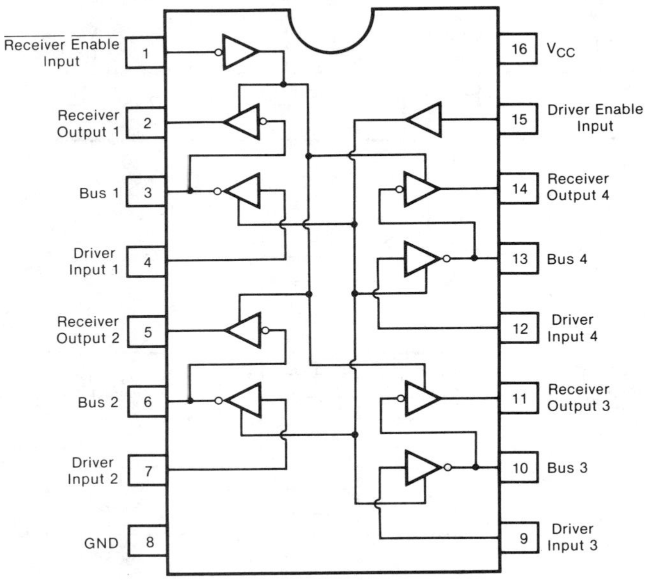


**Figure 4-3.** 6880 pin connections and internal structure.

## The MC6800 Microprocessor

The address bus and bidirectional data bus were discussed in Chapter 3. The hardware aspects of the components in the system are interconnected so that the data are properly routed between them and to any external devices being used. In addition to these two buses, power, and ground, numerous control lines are required. These lines are called the *control bus*. To troubleshoot a microprocessor system, you need to understand the function and characteristic of each control signal. For the technician, this is probably the most important aspect of the microprocessor's operation.

The 6800 utilizes a non-TTL-compatible, two-phase clock to control its operation. The clock signals synchronize the internal operation of the microprocessor as well as the external devices on the bus. The phase 1 clock is on pin 3, and the phase 2 clock is on pin 37. Figure 4-4 shows the two-phase, nonoverlapping clock signal waveforms. The specific times depend on the particular crystal frequency, which may be from 100 kHz up to 1 MHz for the MC6800B, and up to 2 MHz for the MC68B00. Certain components and functions of the system affect the clock requirements. Motorola makes a 16-pin IC, the MC6875, that includes all of the features necessary to control the timing of the 6800 microprocessor system. If, for example, the system contains slow memories, the clock can be stretched by holding the memory ready, pin 6, on the clock chip low. This keeps phase 2 high, which allows data to be transferred until the memory ready input goes high. The MC6875 also has provisions for direct memory access (DMA), dynamic RAM (DRAM) refresh, and a Schmitt trigger input that controls RESET.

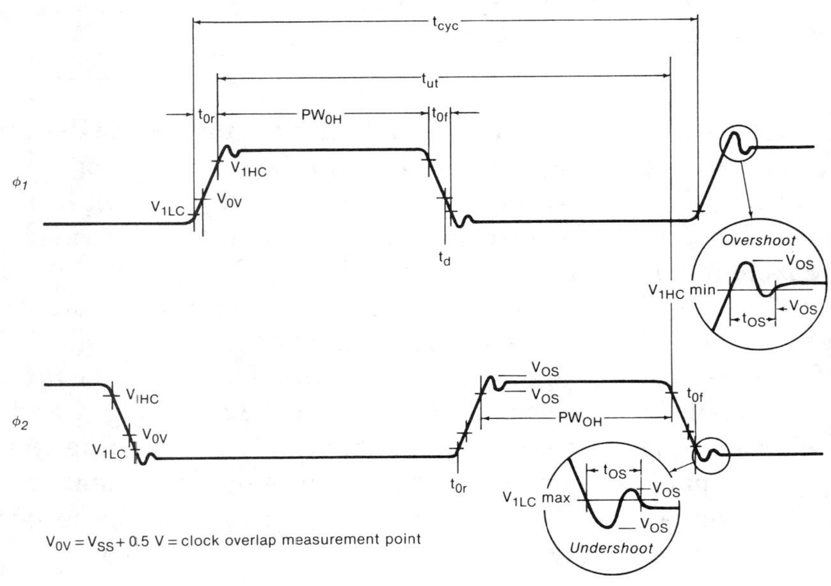


**Figure 4-4.** Two-phase nonoverlapping clock signals.

The R/W output line, pin 34, is used to signal all external devices that the microprocessor is in a read, a write, or a high-impedance state. When the three-state control (TSC), pin 39, is high, the R/W line is put in the high-impedance state.

When the HALT input, pin 2, is high, the microprocessor is running and executing instructions. When it goes low, the microprocessor finishes execution of its current instruction and then halts. In the HALT mode, all three-state lines go to the high-impedance state, and the bus available (BA), pin 7, goes high. Also the VMA, pin 5, goes low. The BA signal is a normally low signal generated by the microprocessor. This signal may also be put in the high state by a wait (WAI) instruction. The VMA is used to indicate to all of the external devices that there is a valid address on the bus when it is high. VMA is not a three-state line, so it must be externally buffered in some applications, such as DMA.

The data bus enable (DBE), pin 36, enables the data bus drivers of the microprocessor when it is in a high state and puts them into the high-impedance condition when in the low state. The input to this control is normally the phase 2 clock. However, when it is desired that another device control the data bus, again such as in DMA, the DBE should be held low.

The interrupt request (IRQ), pin 4, is a level-sensitive input that requests that an interrupt sequence be generated within the microprocessor as soon as it finishes the current instructions that it was executing before it recognized the interrupt request. When the microprocessor recognizes the request, it checks to see if the interrupt mask (I) in the condition code (status) register is not set. If it is not, the microprocessor begins an interrupt sequence by storing away on the stack the contents of the index register (IR), program counter (PC), accumulators (A and B), and condition code register. Next, the microprocessor will respond to the interrupt by setting the interrupt mask bit high so that no further interrupts may occur. At the end of the cycle, the microprocessor is caused to branch to an interrupt service routine in memory.

The nonmaskable interrupt (NMI), pin 6, is activated by a low-going edge on this signal input line. Again the microprocessor will complete its current execution cycle, but the interrupt mask bit has no effect on NMI. Also, the IR, PC, A's, and condition code registers are stored away on the stack. Then, at the end of the cycle, the microprocessor will branch to an NMI service routine.

Inputs IRQ and NMI are hardware interrupt lines that are sampled during phase 2 of the clock cycle and will start their interrupt service routine on the phase 1 following the completion of an instruction.

The final control bus signal is RESET, pin 40, which is used to start the microprocessor initially or reset it from a power-down condition resulting from a power failure. RESET must be held low for at least eight clock periods after $V_{CC}$ reaches 4.75 V. Then, succinctly stated, a restart sequence routine will be branched to in memory. The interrupt bit mask is set, and it must be reset before the microprocessor can be interrupted by IRQ.

When checking out the operation of the microprocesor CPU, the first item of importance is to locate $V_{CC}$, pin 8, and ground (labeled $V_{SS}$), on pins 1 and 21. During troubleshooting, you will probably check $V_{CC}$ again and again, so note its location. The $V_{CC}$ should measure +5 V, and ground should be 0 V. The logic probe should indicate a steady high for $V_{CC}$ and low for ground, but these are best checked with a DVM or oscilloscope.

There are 26 other pins that are easily tested: the 16 address lines (pins 9–25) and the 8 data lines (pins 26–33). A DVM, on the address and data lines, should indicate an unsteady voltage roughly between 2 and 4 V. The logic probe here beats the DVM and shows pulse activity on the buses. You can always obtain the most information about what is on a pin by using an oscilloscope.

Three of the control line names have a line

over the name, which means NOT, or, more correctly, active low. These pins (4, 6, and 40) are normally held high and brought low only when their particular job is to be performed. For the purpose of troubleshooting, these pins should measure +5 V or a logic high with the logic probe. If not, you have a clue to the problem.

The rest of the pins may also be tested similarly. The clock pins (3 and 37) are rectangular waves of different phases. With a voltmeter, they should measure about +2.4 V each. The logic probe will show them as *pulses*. Again, the oscilloscope can be used to measure the detail of these signals (see Chapter 6). Analysis of the functions of the other pins will allow you to determine whether the pin should be high or low in a given situation. Notice that pins 35 and 38 are not connected.

## Read-Only Memory

Recall that a ROM has data permanently written into it during manufacture. Because it is nonvolatile, the ROM is used to hold constant data and tables—that is, the program that runs the microprocessor so it can be started again after power has been removed. The MC6830 is the ROM designed originally for use in the 6800 system. It comes in a 24-pin DIP package, contains 1024× 8 bytes of memory space, and has three-state outputs capable of driving the data bus.

Figure 4-5A shows the pinout of the MC6830A. It has an access time of 500 ns, which is fast enough for any 6800 system. Figure 4-5B shows the bus organization of the 6830 with its 8 data lines, 10 address lines, and 4 chip-select (CS) lines. The active level of the four CS inputs and the memory contents are defined for a specific ROM by the customer. A mask is then made, and many identical ROMs are manufactured.

Although the CPU actually does all of the work, the ROM tells the CPU what to do and when to do it. Since the CPU can only read from ROM, there is no need for a R/W line, and data travel only from ROM on the data bus.

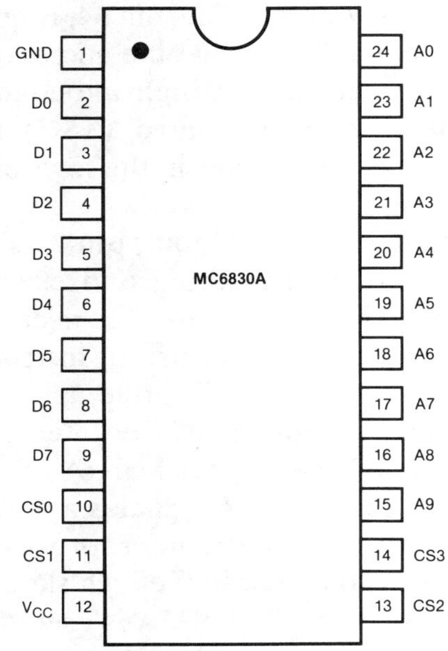

(A) Pinout.

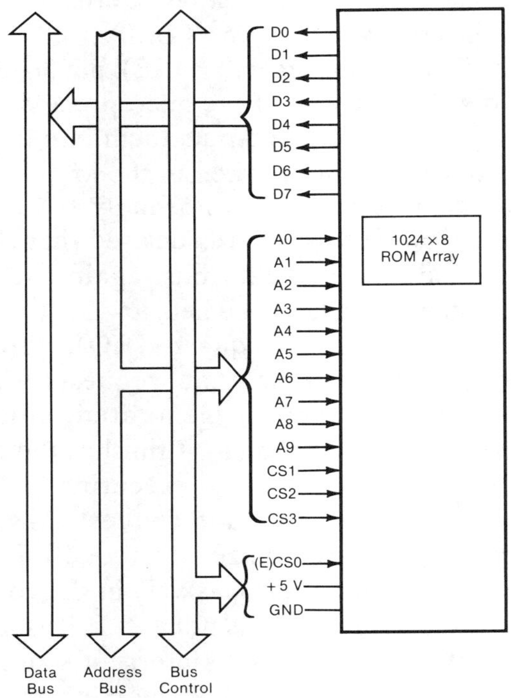

(B) Bus organization.

**Figure 4-5.** MC6830A ROM

The microprocessor system is built so that it goes to the starting address in ROM as soon as it is turned on. This is about all that the CPU knows how to do. When the start address is reached, the ROM takes over and tells the microprocessor what to do next. This permanent program at the starting address is called the *monitor*. Its function is to get the microprocessor underway and oversee the operation. If the ROM should become defective, the microprocessor would, after power-up, go haphazardly from one address to another or else hang up at some address. Besides replacement, the logic analyzer (see Chapter 7) is ideal for reading the contents of a ROM and comparing it with the known good program.

Should you ever have the need to program your own ROM, you would probably use a programmable read-only memory (PROM) or an erasable programmable read-only memory (EPROM). A troubleshooter might use a PROM or an EPROM to make the microprocessor perform as an "exerciser." When the ROM is replaced, the exerciser is programmed to make a few valuable tests on selected components.

## Random-Access Memory

RAM literally means random-access memory. However, both RAMs and ROMs are random access; that is, each memory location is equally accessible. More generally, a RAM is a read/write memory—it can be both read from and written into. Each word in a RAM has two parameters: its address, which locates it within memory, and data, which are stored at that location. In the process to access the contents of the memory location to either read or write data, the first step is that the seven address lines, A6–A0, are used by the microprocessor to select a memory location on the particular chip addressed. If there is more than one RAM chip like that in Figure 4-6 in the system, the six CS pins will be tied to the address bus in such a manner that the microprocessor addresses only one RAM chip at a time.

Notice that the MCM6810 RAM requires a low on four of the CS pins and a high on two of them. The R/W line is the control line discussed when we considered the 6800 microprocessor. To read data from the selected RAM location, the

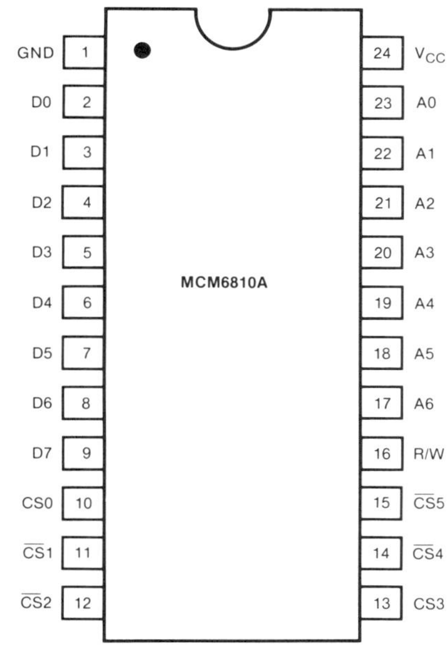


**Figure 4-6.** MCM6810A RAM pinout.

R/W line must be in a high state. To store data in a RAM location, the R/W line must be in a low state. When a particular RAM is not addressed, the RAM's data bus goes to the three-state (high-impedance condition).

RAM and ROM addresses must be selected so that the locations do not overlap; then address decoding is made simply of combinational logic circuits. If a gate in this address decoding circuitry fails, an entire portion of memory, or at least one RAM chip, becomes inaccessible. If the defective part is in ROM, as mentioned before, the computer will lock in a particular condition or operate haphazardly. If in RAM, and the part of RAM disabled is not accessed, the system will function normally. Troubleshooting the address decoding circuitry is just a matter of using the basic techniques for any combinational logic circuit as presented in Chapter 2.

As we will see in Chapter 6, a logic pulser can be used to stimulate the input of gates while the output is monitored with a logic probe or oscilloscope. Blocks of memory that cannot be activated by the enable signal is a condition that you can find by monitoring the CE pins.

Failure of a single memory bit storage location is best found with memory diagnostic software (see Chapter 5). This type of software tests RAM by writing a known code into every storage location and then reading the locations to see whether the code was stored properly. Any difference in what was stored and what was read back is displayed on the monitor screen. Again, as with the ROM, the logic analyzer is the most useful instrument for memory troubleshooting.

The MCM6810 is a static RAM and retains data as long as power is applied. Significantly less power is required, and more memory locations per chip can be obtained by the use of DRAMs. DRAMs store information in the form of a charge on a FET capacitor, which can retain its charge only a few milliseconds, and hence must be refreshed periodically. As mentioned earlier, the refresh control signal can be obtained from the clock chip (MC6875).

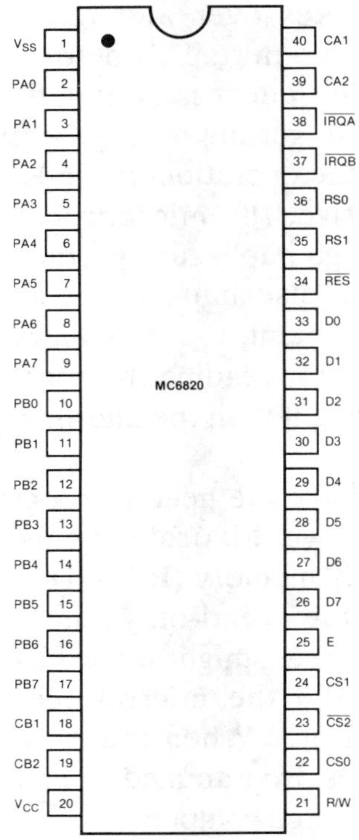


**Figure 4-7.** MC6820 PIA pinout.

## Peripheral Interface Adapter

The PIA provides the universal means of interfacing peripheral equipment to the microprocessor system. The MC6820 is a member of the 6800 family that fills this role. Its pinout is shown in Figure 4-7. Since the PIA can do most routine peripheral control tasks, the microprocessor is freed to handle more important tasks. The 6800 microprocessor system treats the PIA registers as a set of memory locations that must be assigned and addressed as are any other memory locations. The PIA therefore allows the microprocessor to treat a peripheral device as a memory location and, since the PIA has some of its own address decoding, a separate address decoder is not needed in many applications. The MC6821 is an augmented version of the MC6820 that is capable of operation without a clock signal (static operation). Also, it can drive two TTL loads on all A and B side buffers.

The PIA is a dual I/O port unit with two similar sides (A and B). Each side contains three registers, as shown in Figure 4-8. A port is an interconnecting point where data are input and/or output. Two types of I/O ports are in widespread use: parallel and serial. A parallel port allows output or input of a given number of bits at the same time. Data words of 8 bits and 16 bits are commmon. A serial port allows only 1 bit to be input or output at a time.

Each of the peripheral data registers is the interface between the PIA chip and the outside world. Each is 8 bits (1 byte) wide. The data direction register is used by the programmer to define each individual peripheral, though they typically contain 00 or FF so that the entire byte is transferred either in or out. When each bit in this 8-bit register is set to a 1, the corresponding peripheral data line is defined as an output line. When set to 0, the corresponding peripheral data line is defined as an input line. The control register is used to permit the microprocessor to control the operation of the four peripheral control lines CA1, CA2, CB1, and CB2. This register is also used to control the interrupt lines and monitor the status of the interrupt flags. Bit 2 of the control register is used, in conjunction with the register selects, to determine whether the peripheral data register or the data direction register is to be addressed, since they have the same address.

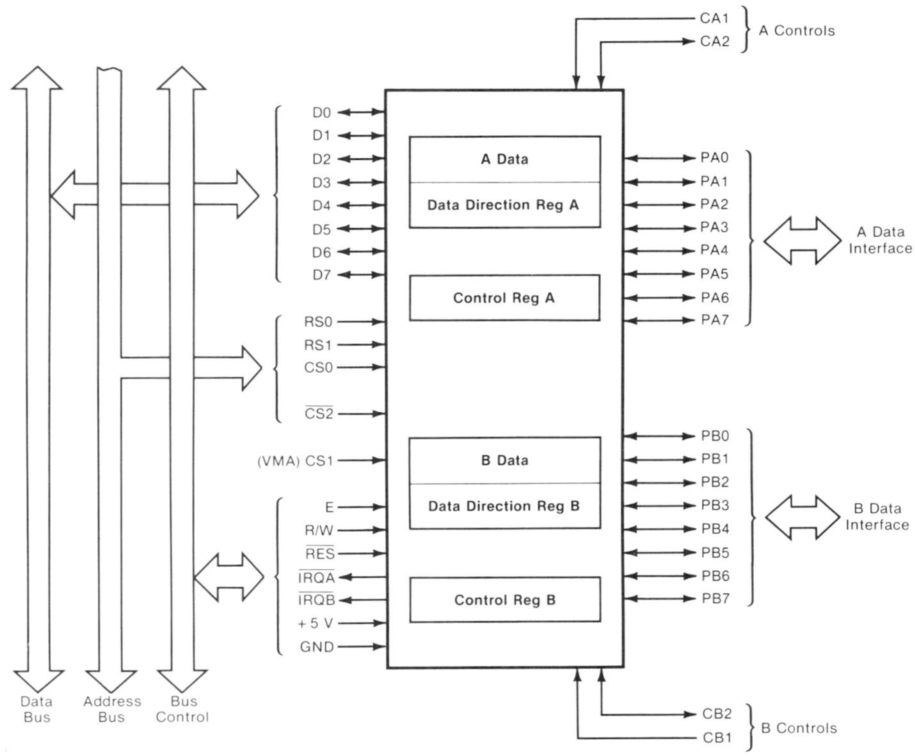


**Figure 4-8.** MC6820 PIA bus interface and registers.

The PIA must always be initialized by a system reset. When the RESET line of the PIA is driven low, it clears all of the registers. Since bit 2 of the control register is 0, the direction register is set to be accessed. The RESET line is normally connected to the microprocessor RESET. Therefore, a RESTART program routine is automatically accessed by the microprocessor whenever RESET is activated. The desired PIA configuration is selected during the execution of this RESTART program routine.

The PIA is a versatile chip and can do many kinds of I/O interfacing tasks. Some microprocessor systems have more than one PIA chip to handle all of the interfacing duties. Since the jobs vary for PIAs with different microprocessor systems, you must know the particular applications so that you can intelligently interpret test results. Even though the pinout numbers are the same on two MC6820 PIAs, the logic probe readings or oscilloscope waveforms are not necessarily the same. In fact, they will usually be quite different. The reason it is so important to realize the differences is due to a popular technique of troubleshooting used in electronic servicing. In a stereo system with only one channel in trouble, comparing readings between corresponding points of the good and the bad channel can often help to locate the trouble source. Be careful when comparing the readings of the same type of PIA chip doing different jobs.

For years the electronic service technician has relied on voltage and resistance charts that are usually a part of the service literature. In troubleshooting, the voltage chart and then the resistance chart are followed. Each voltage point is measured and compared with the chart. When an incorrect voltage is found, it is a clue to the malfunction. Using these charts provides an easy and timesaving method of locating a trouble. They are valuable troubleshooting aids.

Similarly, a logic probe test chart of a microprocessor or other digital system is effective, and the same type of test can be done. These charts are not always available in manufacturers' service literature. However, you could prepare your own. This would be particularly helpful if you are likely to repair the same type of system over and over again.

Troubleshooting the interface circuits therefore requires an understanding of these main chips, their functions, combinational logic circuit elements, and the use of test equipment such as logic probes, pulsers, oscilloscopes, and logic analyzers.

## Asynchronous Communication Interface Adapter

When data are being transferred between components in the microcomputer system, such as the 6800, it is moved a byte at a time by the 8-bit data bus. This type of transmission is parallel transmission, and it is synchronized by the microprocessor system clock. In serial data transmission, each code element (bit) is sent sequentially. The main advantage of serial transmission is that only one pair of wires is required for the transmission line connecting the transmitter and receiver. For long distances, the serial method is obviously preferred, though it is much slower than parallel transmission.

Serial data transmission can be broken down into two separate classes of transmission: synchronous and asynchronous. Synchronous transmission uses a common clock pulse at the transmitting and receiving end. The receiving device is clocked at exactly the same rate as the transmitter by locking onto a unique code in the received bit stream. In asynchronous serial transmission, each character has its own synchronizing information to inform the detector of the first and last bit locations of the character. It can therefore be sent at varying rates. Asynchronous transmission uses start and stop bits to synchronize each character, which allows for the variable interval between the transmitted characters, for example when an operator presses a keyboard character.

In the 6800 microprocessor system, serial data transmission in implemented by the MC6850 ACIA. The ACIA is designed to accept parallel data from the microprocessor system

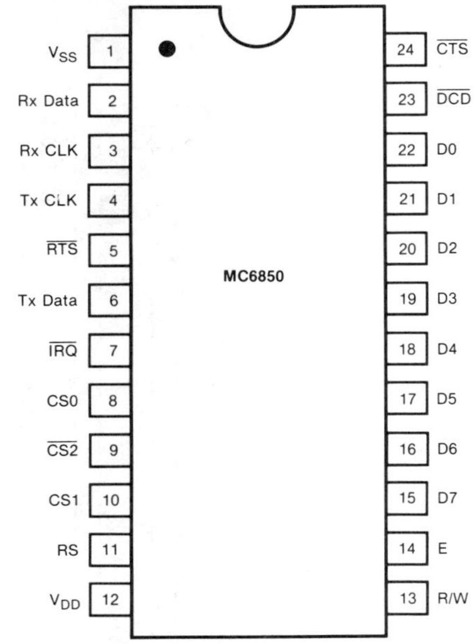


**Figure 4-9.** MC6850 ACIA pinout.

data bus a byte at a time. It then sends the data out serially a bit at a time to an asynchronous serial peripheral device such as a CRT terminal or line printer. Simultaneously, other circuits in the ACIA can accept serial data input to its port and place the bits on the microprocessor bus as parallel bytes. So, the ACIA can function as either a serial-to-parallel converter or a parallel-to-serial converter. The pinout for the MC6850 ACIA 24-pin DIP is shown in Figure 4-9, and its expanded internal block diagram is shown in Figure 4-10. The left side of the block diagram shows the lines that connect to the data, address, and control bus lines of the microprocessor. The right side shows the lines used for the interface with the external peripheral device.

As the block shows, the ACIA has four registers that can be addressed by the microprocessor. It transfers information bidirectionally on the 8-bit data bus just as the RAMs, ROMs, and the PIAs do. Of the four registers, two are read only and two are write only. The read-only registers are the status register (SR) and the received data register (RDR). The write-only registers are the transmit data register (TDR) and the control register (CR).

Also shown in the block diagram are the transmit shift register and the receive shift

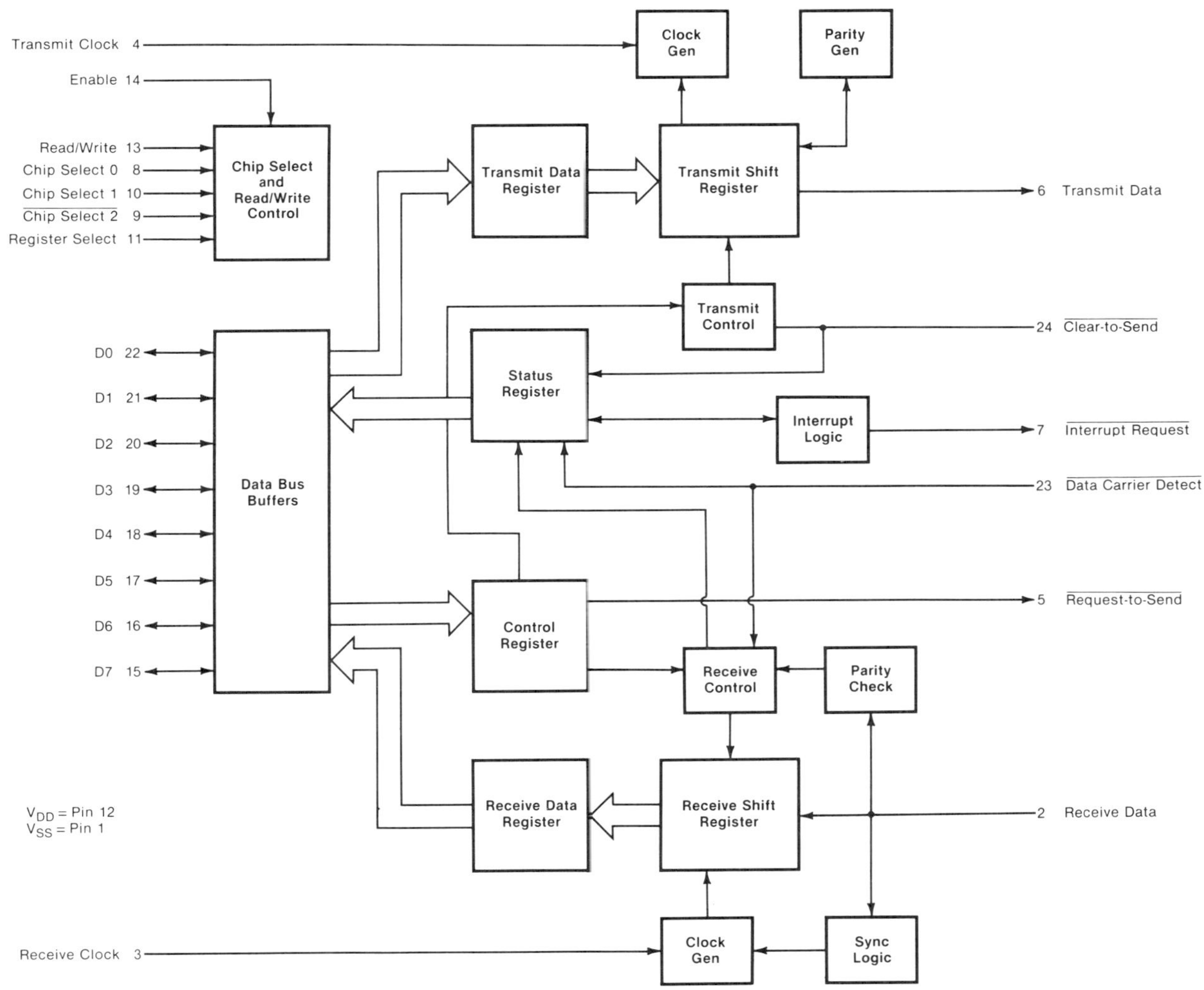


**Figure 4-10.** MC6850 ACIA block diagram.

register. The transmit shift register is loaded in parallel by way of the TDR from the microprocessor via the data bus. Then it is shifted right, which sends the bits out serially on the transmit data (Tx Data) line, pin 6. The receive shift register accepts serial data as input on receive data (Rx Data), pin 2, and sends it out by way of the RDR and on via the data bus to the microprocessor.

The SR monitors the progress of data transmission and reception and sends this information to the microprocessor. It is an 8-bit (read-only) register selected by the CS0, CS1, and CS2 lines, with the register select (RS) line held low and the R/W line high. The 8 bits of the SR, when read by the microprocessor, determine and maintain the internal activities of the ACIA. The operation of the ACIA is shown in some detail in Figure 4-11.

Notice the SR and the CR. The bits of the SR are as follows:

| Bit | Function | Mnemonic |
|---|---|---|
| 0 | Receive Data Register Full | RDRF |
| 1 | Transmit Data Register Empty | TDRE |
| 2 | Data Carrier Detect | DCD |
| 3 | Clear to Send | CTS |
| 4 | Framing Error | FE |
| 5 | Receiver Overrun | OVRN |
| 6 | Parity Error | PE |
| 7 | Interrupt Request | IR |

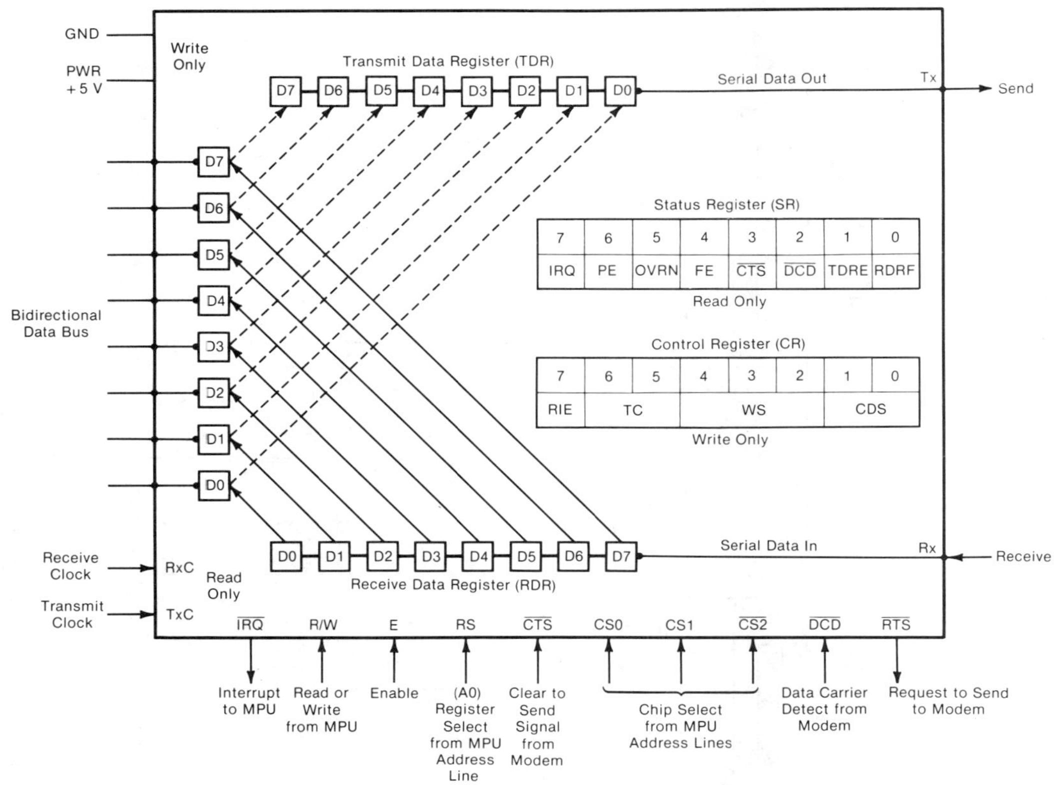


**Figure 4-11.** Operation of the ACIA.

The bits of the CR are as follows:

| Bit | Function | Mnemonic |
|---|---|---|
| 0, 1 | Counter Divide Select Bits | CDS |
| 2, 3, 4 | Word Select Bits | WS |
| 5, 6 | Transmitter Control Bits | TC |
| 7 | Receiver Interrupt Enable | RIE |

The main 24-pin and 40-pin chips are representative of the most complex chips in the microprocessor system. During troubleshooting, a lot of activity takes place around them. The more that you learn about their theory of operation, the better equipped you will be to make input and output tests on them and to make a correct decision regarding the cause of the malfunction.

## Video Display Generator

The MC6847 VDG provides a means of interfacing the MC6800 microprocessor family to a standard color or black-and-white television receiver. It reads data from memory and produces a video signal that allows the generation of alphanumeric or graphic display. The generated video signal can be modulated to either channel 3 or channel 4 by using the compatible MC1372 (TV chroma and video modulator).

The pinout for the MC6847 is shown in Figure 4-12. It is a noninterlaced VDG used in TV games and personal computers. It scans one field of 262 lines 60 times per second. Eight TTL compatible data lines are used to input data from RAM to be processed by the VDG.

When you troubleshoot the VDG, I/O tests will most likely disclose the problem. The inputs may be inspected with a logic probe. The outputs require an oscilloscope, which can also be used on the inputs. A logic probe on the outputs will read haphazardly random highs, lows, and pulses—indications that are meaningless as far as troubleshooting goes. With an oscilloscope,

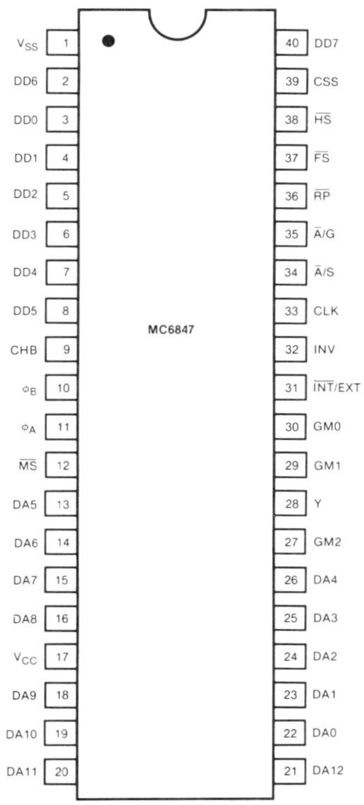


**Figure 4-12.** MC6847 VDG pinout.

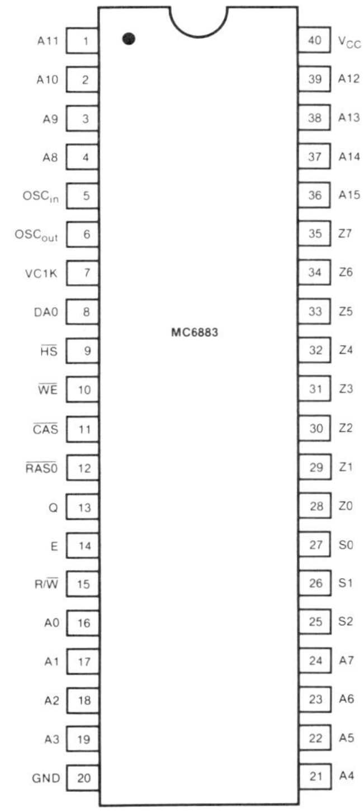


**Figure 4-13.** MC6883 SAM pinout.

the signal can be traced from the Y output of the VDG (pin 28) to the input of the RF modulator or to the video monitor.

## Synchronous Address Multiplexer

The SAM brings together the microprocessor, the VDG, and the DRAM to form the microcomputer display system. The pinout of the MC6883 is shown in Figure 4-13. This one 40-pin chip replaces numerous buffers, gates, and registers, thus significantly reducing the chip count in the microcomputer system. It has four basic functions in the typical microcomputer system:

1. With the aid of an external crystal, it generates all of the timing signals for the microcomputer.
2. It does all of the device selection.
3. It generates video address lines and multiplexes them with the address coming out of the microprocessor.
4. It is a controller for the DRAM.

When an area is a suspect for any malfunction, the best way to test the SAM is to take the schematic of the system, the microprocessor family data manual (for example, Motorola DL-120). Using internal block diagrams of the chips, test the pins for input and output conditions with the logic probe and the oscilloscope.

The more that you know about the main chips, the more proficient a troubleshooter you will be. Remember, the primary object is to locate and repair the defect quickly; that is, get the system up fast without redesigning it.

# Some Additional 6800 Family Chips

We have discussed the most-used I/O support components, the PIA and the ACIA being the most common in the 6800 microprocessor family. Additional I/O devices are available. Consult the manufacturers' literature for details. For rapid reference, the pinout diagrams for these devices are shown in Figure 4-14.

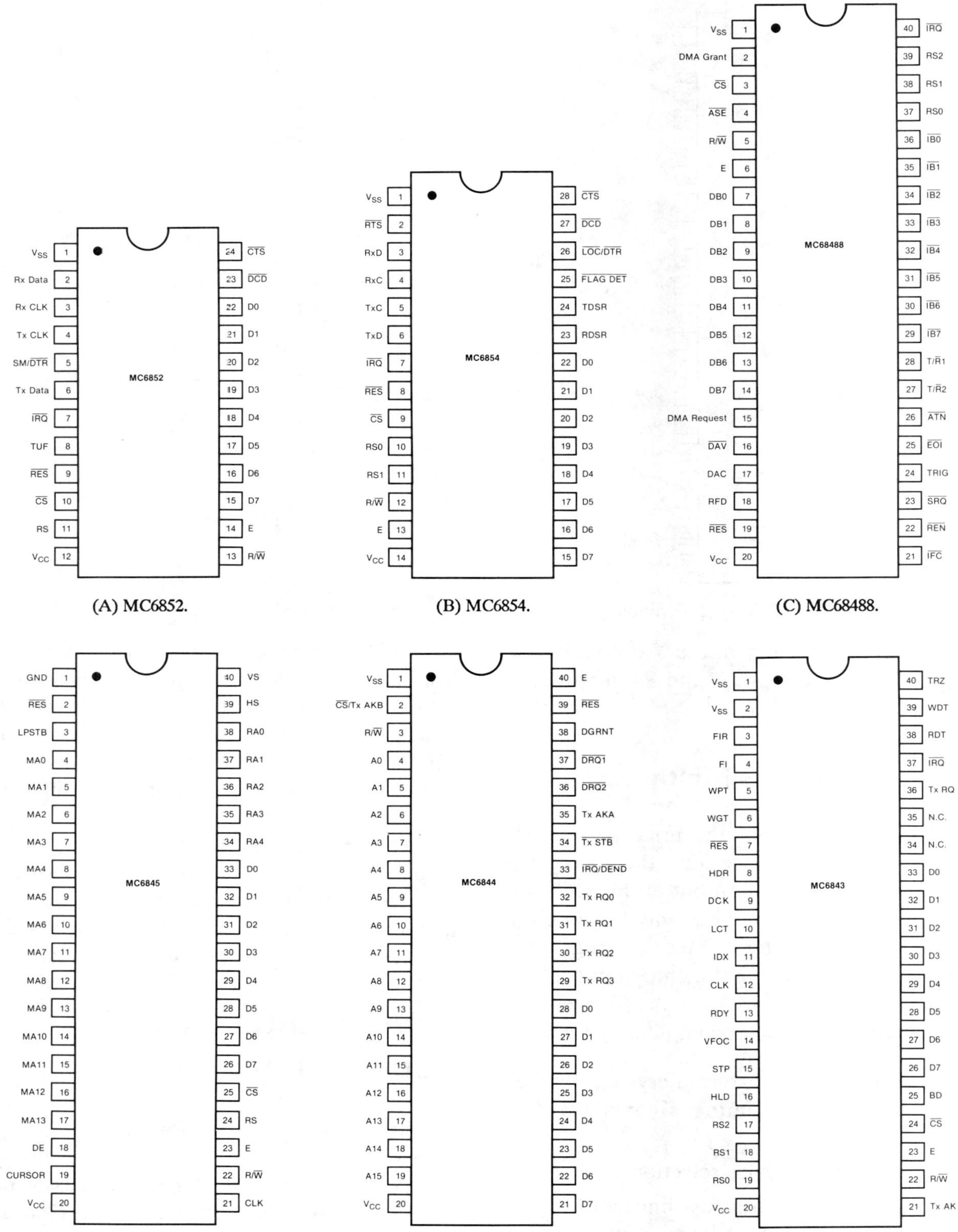


(A) MC6852. (B) MC6854. (C) MC68488.

(D) MC6845. (E) MC6844. (F) MC6843.

**Figure 4-14.** Pinouts of additional 6800 chips.

## Synchronous Serial Data Adapter

The MC6852 synchronous serial data adapter (SSDA) provides a bidirectional serial interface for synchronous data information interchange. It contains interface logic for simultaneously transmitting and receiving standard synchronous communications characters. Typical applications of this device include floppy disk, cassette, or cartridge tape controllers, data communications terminals, and industrial numeric control systems.

## Advanced Data Link Controller

The MC6854 (28-pin) transmits and receives data in a format called a *frame*. All frames start with an opening flag (F) and end with a closing flag (F). Between the opening and closing flags, a frame contains an address field, control field, information (optional), and frame check sequence field. The flag is the unique binary pattern 01111110. It provides the frame boundary and a reference for the position of each field of the frame. The MC6854 advanced data link controller (ADLC) performs the complex microprocessor/data communication link interface for both primary and secondary stations in stand-alone, polling, and loop configurations.

## General-Purpose Interface Adapter

The IEEE-488 instrument bus standard is a bit-parallel, byte-serial bus structure designed for communications to and from intelligent test instruments. With this standard, many instruments may be interconnected, remotely and automatically controlled, or programmed. Data may be taken from, sent to, or transferred between instruments.

The MC68488 chip provides an interface between the IEEE-488 standard instrument bus and the MC6800 microprocessor family. The general-purpose interface bus (GPIB) instrument bus provides a means of controlling and moving data between instruments connected to it. It will automatically handle all handshake protocol needed on the instrument bus.

All inputs to the general-purpose interface adapter (GPIA) are high impedance and TTL compatible. All outputs from the GPIA are also TTL compatible. The interrupt request (IRQ) and the service request (SRQ), however, are open-drain outputs and require that they have an external pull-up resistor (nominal 3.3 kΩ) or be in a wired-OR to the bus.

## CRT Controller

The CRT controller (CRTC) performs the interface between the microprocessor and a raster-scan CRT display. The MC6845 CRTC consists of programmable horizontal and vertical timing generators, programmable cursor logic, light pen capture register, and control circuitry for interface to a microprocessor bus. The internal CRTC registers are programmed by the microprocessor through the data bus D0–D7 and the control bus signals R/W, CS, RS, and E. In a graphic system, both the memory addresses and the row addresses continue to run during retrace, thus allowing the CRTC to provide the refresh addresses required to refresh DRAM.

## Floppy Disk Controller

The register section on this MC6843 floppy disk controller (FDC) consists of 12 user-accessible registers used for controlling a floppy disk drive. All 12 are connected by the internal data bus to allow the processor access to them. This MC6843 FDC is a single-density controller, which is IBM compatible. The FDC can interface a wide range of drives with a minimum of external hardware. Multiple drives can be controlled by using external multiplexing instead of additional FDCs.

## Direct Memory Access Controller

Direct memory access was discussed in Chapter 3. Recall that the transfer of data from a disk to memory, or vice versa, may be achieved most rapidly if it is done directly by hardware rather than by using software to make the transfer through fetch and execute cycles of the microprocessor. The direct memory address controller (DMAC) is a dedicated processor of data used for fast and efficient transfer of blocks of data. The MC6844 performs this function of transferring data directly between memory and peripheral device controllers. It directly transfers the data by controlling the address and data buses instead of controlling the microprocessor and having it do the data manipulation. The DMAC works as an interface with the FDC and ADLC.

The MC6800 family is a popular 8-bit microprocessor family in industrial use. Though not exhaustive, this discussion should serve to acquaint the troubleshooter with the techniques for servicing any similar microprocessor family system. Let's look now at an example of a popular 16-bit microprocessor and its family of chips.

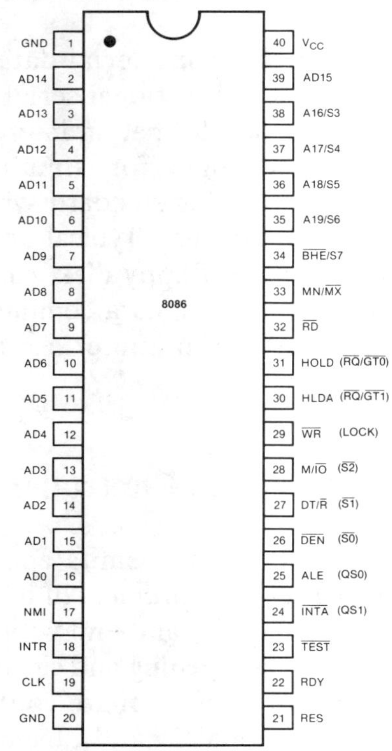


**Figure 4-15.** 8086 pinout.

## Introduction to the 8086 Microprocessor System

The Intel 8086 is a full 16-bit microprocessor; that is, it has a 16-bit data line internally and brought out to its pins. It has a 20-line address bus, 16 of which are also the multiplexed data bus lines. The pinout for this microprocessor is repeated in Figure 4-15. Do not be overwhelmed by all the different mnemonic pin names. Many of these will become clear as we discuss the microprocessor and some peripheral chips.

The internal functions of the 8086 microprocessor are logically divided into two separate processing units: the bus interface unit (BIU) and the execution unit (EU). The BIU is responsible for performing the bus operations, such as instruction fetching, reading and writing information from/to memory, and transferring data between the microprocessor and peripherals. The EU is responsible for executing instructions. Though these units can interact directly for the most part, they perform as separate asynchronous operational processors.

The 8086 is equipped with a strap pin (MN/$\overline{\text{MX}}$) that defines the system configuration as minimum mode or maximum mode operation. A logic high (1) on this pin 33 selects the minimum mode, which is typically a small system utilizing a single microprocessor. If a logic low (0) is applied to this pin, the internal architecture is configured into the maximum mode of operation. The maximum mode is used for a multiple-processor system. As shown in Figure 4-15, the pin functions specified in parentheses are those that pertain to a maximum mode system. We will briefly consider only the minimum mode.

The 8086 in the minimum mode provides the necessary control signals to implement all the interfaces, both memory and I/O. This configuration is shown in Figure 4-16. The signals can be broken into five basic groups:

1. The address/data bus, which, as mentioned above, consists of 20 address lines A0–A19. The 16 data bus lines D0–D15 are multiplexed with address lines A0–A15, respectively.
2. The four most significant address lines, A19–A16, with the four most significant status signals, S6–S3 (S2–S0 are used in maximum mode).
3. The control signals are provided to support and control functions between the microprocessor and memory and/or I/O.
4. The interrupt interface signals are used by a peripheral device to signal the microprocessor that it needs to be serviced.
5. The DMA interface is used when an external device wants to take control of the system bus. It signals the microprocessor by switching signal input pin 31 (HOLD) to the logic high state. When in this state, the microprocessor's 20 address lines, as well as BHE, M/IO, DT/R, RD, WR, DEN, and INTR, are all put in the high-impedance three-state condition. The HLDA is an output signal pin that goes to a logic high to signal external devices that the 8086 is in this operational state.

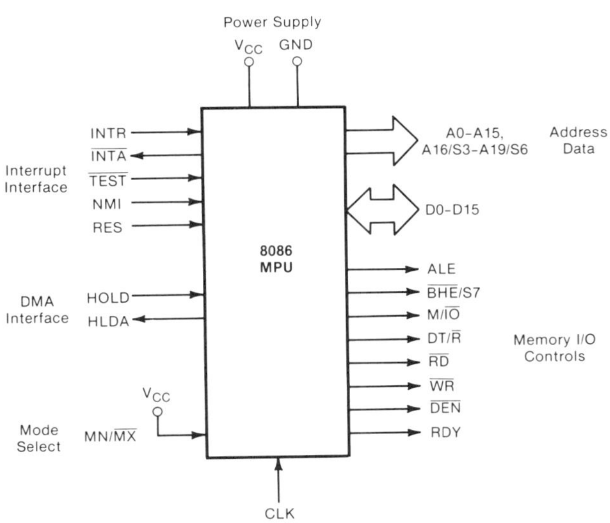


**Figure 4-16.** Block diagram of the minimum-mode 8086 microprocessor.

## A Minimum-Mode 8086 System: Support Chips

Let's examine briefly some of the hardware interfaces of the 8086. The pinout diagrams are shown in Figure 4-17 for several of the common interface chips. Additional information may be found in Intel's *Microprocessor and Peripherals Handbook.*

### Clock Generator and Driver

The 8284 is a single-chip clock generator/driver for the 8086 microprocessor. It uses an external crystal from which the basic operating frequency of the system is derived.

### Octal Latch

The 8282 is an 8-bit bipolar latch with three-state output buffers. It can be used to implement latches, buffers, or multiplexers. The STB is an input control pulse that strobes data into the data latches. The data on the data input (DI) pins are latched at the high-to-low transistion of the STB. The output enable (OE) is an input that, when active low, enables the data to the data output (DO) pins. The 8283 is identical except the output is inverted ($\overline{DO}$), whereas the 8282 is not.

### Octal Bus Transceiver

The 8286 is an 8-bit transceiver with high-impedance outputs. With transmit (T), pin 11, active high and OE, pin 9, active low, data at the A0–A7 pins are driven into the B0–B7 pins. Alternatively, with T in active low and OE in active high, data at the B0–B7 pins are driven onto the A0–A7 pins. There is also an inverting version of this chip, the 8287.

### DRAM Controller

The 8203 is a DRAM system controller that can provide all the signals necessary to control 64K (2164) and 16K (2117, 2118) dynamic memories. Pin 35 is labeled 16K/64K, which selects the

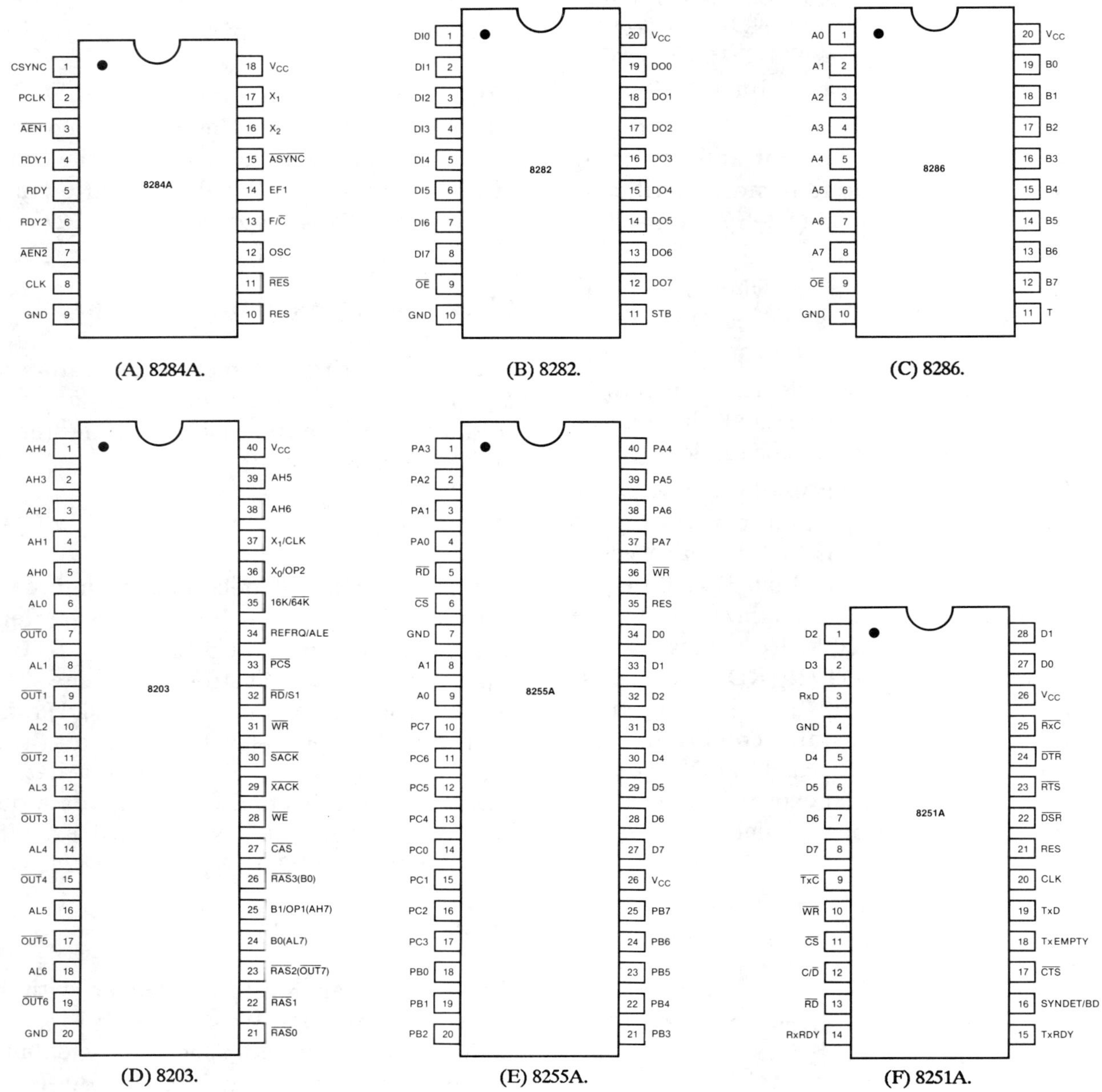


**Figure 4-17.** Pinouts of Intel's microprocessors.

mode or size of memory. The 8202 is just like the 8203 in the 16K mode. It does not have a 64K mode, so its pin 35 has no connection (NC).

## Programmable Peripheral Interface

The 8255A is a peripheral chip that permits easy implementation of parallel I/O with the 8086 microprocessor. The functional configuration is programmed by the system software so that normally no external logic is necessary to interface peripheral devices. A data bus register receives data from the microprocessor via the bidirectional data bus. Then, as selected by software, 3-byte-wide I/O ports called Port A, Port B, and Port C can be configured for input or output operation. This gives a total of 24 I/O lines.

### Programmable Communication Interface

The 8251A programmable communication interface (PCI) can be used to implement a serial data communication interface such as an RS-232C port. It is the type of interface used to connect a CRT or printer to a microcomputer system. Like most of the 8086 family support chips, its functional configuration is programmed by the system's software. A set of control words must be sent out by the CPU to initialize the 8251A to support the desired communications format. These control words will program the baud rate, character length, number of stop bits, synchronous or nonsynchronous operation, even/odd/off parity, and so on. Once programmed, the 8251A is ready to perform its communications functions.

# 5

# Software and the Diagnostic Program

In this chapter, software is examined with its addressing modes, the instruction set, and the analysis and writing of assembly language programs. With specific diagnostic software, the technician can quickly test hardware and identify the problem. We will compare troubleshooting techniques that use functional testing with those that use diagnostic software. We will see how plug-in diagnostic PROMs and high-level-language diagnostic diskettes are utilized in digital troubleshooting.

## Microcomputer Software

Software is what makes microprocessor technology different from conventional electrical and mechanical technology. In the fundamental sense, software is a set of instructions that tells the hardware what to do. The hardware is the microprocessor and any electrical or mechanical devices connected to it. In conventional technology, the arrangement of the components determines what the machine will do. In a microprocessor system, it is not always possible to tell from the physical arrangement exactly what the system does. Its operation is determined by software.

At least two fundamental characteristics distinguish software from hardware:

1. Software is easier to manipulate and modify than hardware. To change a piece of software usually involves a few keystrokes at the keyboard. Changing a hardware layout might require such things as disassembly of the system, unsoldering parts or wires, replacing chips or boards, and reassembling.

2. The physical form of the software can be changed without altering its essential nature. For example, the same piece of software could exist on magnetic disk or tape, in semiconductor memory, as hard copy printed out, or be simply displayed on a screen.

## Software Development Process

The major effect of design and development for a microprocessor system goes into software. The software designer spends much time typing information at a keyboard and looking at results displayed on a screen. Using a high-level language such as BASIC, FORTRAN, Pascal, or C allows the software designer to make strategic decisions about what the system will do. The designer might use a compiler, which translates the high-level-language programs into the detailed machine instructions required by the microprocessor. It is the compiler, a software utility, that determines the tactics employed, such as selecting memory addresses, storage allocation, and other housekeeping functions that have to be performed thousands of times a second.

Programming too early in the software development process—that is before a software design has been worked out—produces an incoherent, badly structured, and inefficient design. Software design consists of turning the specification of each function to be performed by the microprocessor into a precise step-by-step procedure called an *algorithm*. This is not yet programming, which occurs at a more detailed level. Programming involves turning a software design into source program code, following the syntax rules for a particular programming language. The amount of work involved depends on the programming language selected for implementation. Selection of the language depends on what the program will be used for, the development facilities available, and the skill of the designer.

### Configuration and Linking

Most software systems are written as several small packages, which are easier to manage than a large piece of software and take less time to translate. Therefore, most operating systems permit the creation of collections of subroutines or subprograms called *libraries*, which can be meshed into one complete software system. A system library for general use is normally available, and users can create their own libraries.

Configuration is the process of selecting the pieces of software required for an application and linking the pieces together. Joining the pieces (software packages) is done automatically by a software tool called a *linkage editor* or *linker*. Another software tool used for transferring the program to be executed into memory is called a *loader*. Sometimes the linker and loader are combined into a single program.

### Debugging

In the software development process, *debugging* is the discovery and correction of programming errors. Few programs run correctly the first time. Debugging tools such as monitors and editors, debugging software packages, software simulators, emulators, breakpoints, and logic analyzers (discussed in Chapter 7) make the process easier. Sometimes program debugging and testing is referred to as verification and validation, respectively. *Verification* ensures that the program performs its intended function, and *validation* ensures that the program produces the correct results for a set of test data.

A *monitor* is a simple operating system that allows the user to enter or change programs and data, to run programs, and to observe the status of the various sections of the microprocessor or computer system.

An *editor* is a program that manipulates text material and allows the user to make corrections, additions, deletions, and other changes.

A debugging software program or *debugger* is a program that helps find and correct errors in the program. An example is EXBUG, a program written to control the Motorola microprocessor development system called the EXORciser. Up to eight breakpoints can be set by EXBUG.

A *breakpoint* is a user-specified location at which program execution is to temporarily end. Once the breakpoints are set, the program is started and runs until it encounters one of the breakpoint addresses. It then enters a debug routine, which displays the contents of the microprocessor registers as stored after the last instruction was executed. The operator can then examine these plus memory location contents before continuing to trace through the program. Multiple breakpoints can be set at one time in order to catch the program no matter which branch it takes. The breakpoint function is one of the most useful tools for rapidly finding problems in a system.

A *software simulator* is a computer program that simulates the execution of programs on another computer. It usually runs on a large computer and simulates the working of a smaller computer system that may not be available or even built. A simulator provides a powerful debugging facility that can be used to check out completely the logic of a program. However, it does not prove that the software will work correctly with the actual system hardware. The critical stage of software/hardware integration is best handled by *emulation*, the subject of Chapter 8. Once the system is working in emulation, it is ready for the final stages of development.

### Evaluation and Testing

Once the software code has been debugged and is running on prototype hardware, an evaluation of performance should be made. This performance analysis is usually an automated data collection, reduction, and processing technique. It can be done by a computer or a logic analyzer.

Program testing is essentially a later stage and certainly closely related to debugging. It is a matter of validating the program by trying it out on a suitable set of test cases. Many microcomputer programs depend on real-time inputs from peripherals, which are difficult to control or simulate. Many of the debugging tools mentioned are useful for this task.

### Documentation

The documentation stage of software development is often slighted or overlooked. However, it can be an indispensable part of debugging and testing as well as being valuable in troubleshooting, maintenance, and redesign of the software/hardware system. A properly documented program can be followed by someone else; an undocumented program can be so time-consuming that it is usually easier to start over.

Documentation includes several forms of descriptive material that allow program users to understand the program operation and logic. The forms of documentation that should be included are flowcharts (discussed in Chapter 2), source and object program listings, descriptive comments, memory maps, and circuit diagrams.

As with hardware, the design, implementation, and maintenance of the software may be done by several people and thus must be well organized and developed. A few simple procedures should help to make the program documentation clear to anyone who needs to use it. In general, each page of the program should contain the name of the program, the page number, and the name of the programmer. Each program should have a version number and an abstract describing what the program is supposed to do.

Comments are a basic part of program documentation, although they should not be a substitute for proper programming style. They should be used generously to describe the specific functions of an instruction or group of instructions. They should not merely repeat the instruction mnemonics or meaning, but they should explain the purpose of the instruction or sequence of instructions. Documentation should be concise and limited to important points. Too many or too few comments can make the program difficult to follow. Comments provided in a systematic manner can be helpful in all stages of software development and maintenance.

A *memory map* is a list of memory assignments made in the program. These prevent program routines from interferring with each other and help determine the amount of memory needed. They also help in finding the specific location of parameters, subroutines, tables, and temporary storage.

The importance of clear well-illustrated, accurate, and concise documentation cannot be overemphasized. Documenting the software is as essential as the hardware circuit diagrams are for development and maintenance of the computer.

## Programming Languages

The amount of work involved in turning a software design into a source program code depends, for the most part, on the programming language selected for implementation. Turning a software design into a language such as Pascal or BASIC should involve little more than formalizing it and writing it to conform with the syntax rules. Along with C, Pascal and BASIC are high-level languages that handle much of the routine work of automatically translating the design into machine terms by a software utility, the compiler. BASIC is designed for simplicity. Though it is not as powerful as Pascal, it is ideal for experimental and low-volume designs. The most powerful and time-consuming language to write, and the most difficult in which to develop skill and discipline, is assembly language.

### Machine Language Programming

The program that runs the microprocessor must be stored in successive memory locations in binary form. The earliest computers were programmed directly in machine code—that is, binary digits. Today, three languages can be used to write a program for a microcomputer: machine language, assembly language, and high-level languages.

Each instruction in the computer is represented by a unique pattern of bits that make a byte or a word of program code. Manufacturers define the machine language their microprocessors understand and the set of instructions it uses. As a general rule, two different microprocessors will have different machine languages. For example, the ADD instructions for three microprocessors are

| Microprocessor | Binary |
|---|---|
| Motorola 6800 | 10111011 |
| MOS Technology 6502 | 01101101 |
| Intel 8080 | 10000110 |

This means that the machine language program written for one microprocessor will usually not work on another. There are a few exceptions to this. The Intel 8088 and 8086, 8-bit and 16-bit microprocessors, respectively, are fully software compatible. Zilog's Z-80 instruction set contains all of Intel's 8080A instruction set as well as several enhancements.

The object of any programming task is to convert a particular sequence of instructions into machine language codes and load them into the computer's memory. For this reason, the machine language program is referred to as the *object codes* or *object program*. Programming in machine language is extremely tedious and prone to error. Machine language can be improved by using octal or hexadecimal instead of binary for the codes. Hexadecimal is standard in work with microprocessors. Writing a program in hexadecimal is much simpler and less error-prone than binary is, because each entry requires only two digits instead of eight. In the previous example for ADD, the hexadecimal representations would be

| Microprocessor | Hexadecimal |
|---|---|
| Motorola 6800 | BB |
| MOS Technology 6502 | 6D |
| Intel 8080 | 86 |

### Assembly Language Programming

As mentioned previously, machine language programming is the lowest level of programming because the programmer is communicating with the computer at the computer's basic language. Sometimes a troubleshooter has no choice but to communicate at this level. However, much frustration and many difficulties can be overcome by using the next higher-level language, namely, assembly language, which uses mnemonics instead of binary or hexadecimal representation for the instruction codes. The assembly language program is also referred to as a *source program*. It is then converted into a machine object program. This conversion process is called *assembling the program*; the software that does this is referred to as an *assembler program*.

Assembly language has at least two distinct advantages over machine language. First, each assembly language instruction is represented by a mnemonic (an acronym that is easily memorized because what it stands for is usually obvious). Here are a few examples of assembly language instructions and their respective mnemonics:

| **Instruction** | **Mnemonic** |
|---|---|
| Load Accumulator | LDA |
| Add | ADD |
| Subtract | SUB |
| Store Accumulator | STA |
| Halt | HLT |

The second advantage of assembly language is that character strings represent addresses called *identifiers*. A typical assembly language instruction

ADD AX, VOUT

adds the contents of the memory location associated with the identifiers VOUT to the register AX.

## Writing a Program

To write a program requires only a pencil and paper, and a lot of skill, which can be developed with good practice and experience. There are many good books and courses that teach programming in any language including assembly.

One of the first steps to writing a program is to generate a flowchart of the activities that the program is to accomplish. The flowchart for troubleshooting is defined and discussed in Chapter 2. After the flowchart is generated, writing the program is relatively simple. Assembly language uses the unique one-, two-, or three-letter combinations (mnemonics) as the instructions to be performed. Each instruction in the source code (program) corresponds to one assembly language instruction. The statement must specify the operation to be performed and the data operands to be processed. Therefore, an instruction has two parts. The *op code* (operation code) is the part that identifies the particular operation to be performed (e.g., ADD, ROL, and MOV for add, rotate contents of accumulator left, and move, respectively). The *operand(s)* portion of the instruction identifies the data to be processed by the microprocessor as it carries out the operation specified by the op code. An example of an assembly language instruction for an 8088 or 8086 microprocessor is

ADD AX, BX

which adds the contents of registers A and B and puts the results back in A. This example shows one type of instruction and format used for one microprocessor. On the 6502, the same instruction looks like this:

ADC $106A

The addition operation is indicated by the ADC; the operand $106A is hexadecimal. This operand is added to the carry flag and then to the contents of the accumulator. The results are then stored in the accumulator.

Hence, the instructions can be stated in different ways and with different formats. However, every microprocessor has certain types or classifications, though even these groupings may be expanded or overlapped. An example of a set of convenient labels for boxes might be status control, data transfer, data manipulation, program control, and memory reference.

An assembly language program is easier to write and easier to follow than a machine language program, but it eventually has to be converted into machine language before it can be placed in memory and executed. The process of converting an assembly language (source) program into a machine language (object) program is called assembling the source program. It can be hand assembled, or it can be done by the computer. For the computer to do this, it must have a special program, called an *assembler*, stored in its memory. The assembler program can be in ROM, or it can be loaded into RAM from a mass storage device.

Although assembly language is much more convenient than machine language, it still requires that the microcomputer programmer have a fairly extensive knowledge of its internal structure as well as a complete set of instruction mnemonics. In both machine language and assembly language, the programmer must write an instruction for each step the computer is to perform. Also, with the exceptions noted, both of these languages are machine independent; that is, they may be run only on the microprocessor for which they are written. To overcome these difficulties, designers have developed a number of high-level languages among the more popular of which are FORTRAN, COBOL, BASIC, and Pascal. These high-level languages make it possible to communicate with a computer using English words and mathematical symbols without the need for detailed knowledge of the computer's internal architecture or instruction set.

Because we will be working mostly with hardware and its related signals in troubleshooting a microprocessor-based system, we will mostly use assembly language. So, now let's look at ways that instructions for the microprocessor may be classified.

## Addressing Modes

The addressing mode refers to the method by which an instruction addresses its operand—in other words, the method by which a processor can access data. In assembly language statements, the addressing mode is indicated in the instruction. The 6502 microprocessor has 11 different addressing modes, whereas the 8086 has 9. For example, the STA (store accumulator) instruction on the 6502 can use any of seven different address modes. This means that seven different op codes are required to specify the seven different STA instructions. The following addressing modes are the most widely used:

1. Immediate addressing. In this mode, a source operand is part of the instruction. It is loaded into the accumulator, where it is not necessary to specify an operand address. The data (operand) are specified immediately following the op code.
2. Direct addressing. This means that the actual address is part of the instruction. It requires 2 bytes of memory. The first byte is the op code of the instruction; the second byte is the address of the operand, not the operand itself. For example, on the 6800 microprocessor we might use the instruction ADD #45 or ADD 45. ADD #45 uses immediate addressing; the op code 8B 45 would load the data 45 in the accumulator A. The second instruction uses direct addressing; the op code 9B 45 would load the data at memory address 45 into accumulator A.
3. Indirect addressing. In this mode of programming, the address (location) of the address is part of the instruction, rather than the address itself. This mode is slower than direct addressing because the CPU must perform an extra memory access. The CPU must first fetch the instruction containing the indirect address from memory, then use the address to transfer data. An advantage of indirect addressing is that a program can save a result in memory for later use with the instruction. It is particularly convenient for subroutines because it allows the same subroutine to process data in many different areas of memory.
4. Indexed addressing. This is a powerful addressing mode that makes it easy to handle tables or blocks of data in memory. It means that the CPU adds the contents of an index register to the address supplied with the instruction in order to find the effective address. In general, every instruction that involves an operand in

memory can use the indexed addressing mode. When indexed addressing is being used, the address of the operand is determined by adding the offset address contained in the instruction (usually an unsigned 8-bit binary number stated in hexadecimal) to the address in the index register (a 16-bit number). The sum, a 16-bit number, becomes the address of the operand. It is important to remember that this does not change the number in the index in any way.

5. Relative addressing. This addressing mode uses a technique called *branching*. A branch instruction has an address associated with it that indicates the location of the next instruction to be executed rather than the address of the data. In relative addressing, a number following the op code must be added to the program counter to form the new address. This allows us to move programs to different places in memory. Such programs are said to be *relocatable*. Relocatable programs are useful because they can occupy any area of memory; they can be put where they will not interfere with data or other programs.
6. Register direct addressing. This mode is faster than direct addressing. In fact, it is the fastest mode available in most microcomputers. The address following the op code is that of a register instead of a memory location. Some microprocessors must use register direct addressing when performing arithmetic and logical operations. However, the power of this mode is that the program can use the same data many times. For example, if we want to add the contents of a memory location to 100 different numbers, then the register need only be loaded once. The memory does not have to be accessed 100 times. This obviously saves many machine cycles. One caution: The number of registers is limited in any microprocessor, so they must be allocated carefully.
7. Register indirect addressing. The only difference between this addressing mode and indirect addressing is that the address is that of a register instead of a memory location. Register indirect addressing is faster than ordinary indirect addressing because the CPU need not fetch the effective address from memory.
8. Base addressing. In the base addressing mode, the physical address is obtained by adding the contents of a memory location or register to a specified number called a *displacement*. Base addressing may be used in conjunction with direct or indirect addressing.
9. Port addressing mode. This mode is used in conjunction with the IN and OUT instructions to access input and output ports on, for example, the 8086 microprocessor. The IN instruction copies a byte or word from a specified port to an accumulator, whereas the OUT instruction copies a byte or word from an accumulator to a specified port. For ports in the I/O address space, only the direct addressing mode using a specific register (DX) is available.

Most of these addressing modes are available, possibly in modified form, in the more common microprocessors. It is improbable that all of the modes mentioned here are utilized on any specific microprocessor, though each address method has its own special uses and advantages. Few programs use only one addressing mode throughout.

## Microprocessor Instruction Sets

In this section, we consider some of the op codes that determine what the particular microcomputer can do in a single instruction cycle. We divide the instructions into several categories and discuss representative instructions of various microcomputers in each category.

The total number of possible op codes for any microprocessor is many more than the number actually used to represent different available operations. The total number of possible op codes for the popular 8-bit processors is $2^8$, or 256. However, the number of available types of instructions for the 6502 is 56, the 6800 has 72 different instructions, and the 8086 has 117 basic instructions. The number of operands and addressing modes with these instructions further expand the instruction set into the additional instructions that can be executed at the machine code level. For example, the 6502 can run the 56 types of instructions in various addressing modes to give a total of 145 usable instructions. The 8086 is also expandable into numerous other individual instructions. For example, the basic MOV instruction expands into 28 different individual machine level instructions.

All of these instructions may be broken into groups and categorized in various ways. For our purposes here, they will be grouped according to the following functionally related operations:

- Data manipulation instructions
- Data transfer instructions
- Program manipulation instructions
- Status management instructions

*Data Manipulation Instructions*—Data manipulation instructions include mainly the instructions that use the microprocessor's ALU. Most of the 8-bit microprocessors have only basic addition and subtraction instructions, whereas newer 16-bit units like the 8086 have multiplication and division operations. The status that results from the execution of arithmetic operations is recorded in the processor status register bits (flags). The common logic instructions that are performed by the microprocessor's ALU are the AND, OR, Exclusive-OR, and NOT. Remember that the AND operation is the most-used logical operation. It is used for masking (that is, examining) parts of a data word.

There are two basic shift operations: logical shift and arithmetic shift. Both can be to the left or to the right. The LOGICAL SHIFT instruction shifts the data and clears the empty positions. The ARITHMETIC SHIFT LEFT multiplies a two's complement number by 2, whereas an ARITHMETIC SHIFT RIGHT divides the number by 2. When an ARITHMETIC SHIFT RIGHT is executed, the number in the affected register is shifted right one position. Figure 5-1 shows the results of arithmetic shifts left and right. Note that 0s are shifted into the right end of the operand, and bit 14 on the left end (the sign bit is unchanged) is shifted into the carry CF bit position. The previous CF bit is shifted out and lost. The SHIFT RIGHT instruction similarly shifts bits to the right, but it does not automatically insert 0s from the left. Instead, it repeatedly extends the most significant bit (MSB) and shifts the least significant bit (LSB) into the CF bit position. This shift may be on a register in the CPU or, in some microprocessors, in a memory location.

Another data manipulation instruction that is similar to the SHIFT is the ROTATE. It also has right and left directions. This instruction allows the shift of the contents of a register or a memory location without losing bits of data. When the ROTATE LEFT instruction is executed in the 6800 microprocessor, either of its two accumulators may be used. The instruction ROLA rotates in the data in the A accumulator clockwise. See Figure 5-2.

An accumulator is shifted into the carry register while the accumulator contents are shifted to the left. The carry bit is shifted into the LSB position in the accumulator. When this instruction is executed, the accumulator and the carry bit form a 9-bit circulating register. This instruction is useful for implementing multiplication and division on microcomputers that lack explicit instructions for these operations.

Another type of data manipulation instruction is a comparison operation. This operation allows us to compare or test data without changing it. It sets the processor status bits (conditional flags) for later use in another operation. It does not actually store a result anywhere. Examples of this type of instruction are CMP (compare), BIT TEST, TEST, and SCAN. The SCAN instruction is used in text editing to search a string of characters for a particular pattern. A flag is set if the pattern is located and the search is successfully completed.

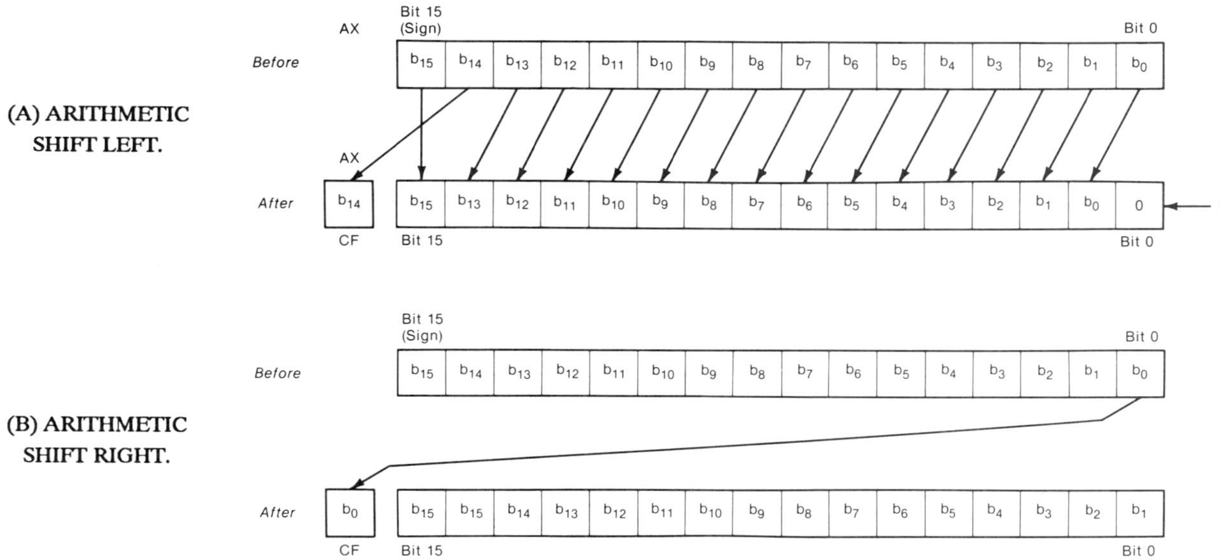


**Figure 5-1.** An ARITHMETIC SHIFT instruction.

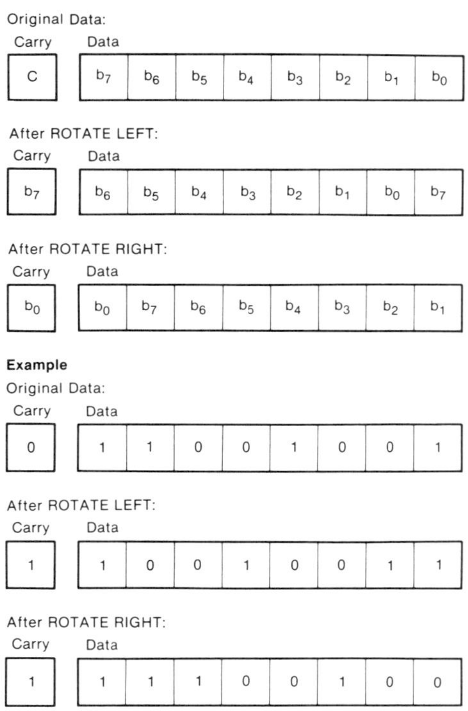


**Figure 5-2.** The ROTATE instructions.

*Data Transfer Instructions*—Data transfer instructions move data from one place in the microcomputer to another without actually making any changes in the data. Instructions such as READ, WRITE, LOAD, and STORE are the most common examples of data transfer instructions. READ and WRITE are I/O instructions, and LOAD and STORE transfer data between registers and memory locations. Data are also moved from one register in the CPU to another register.

Many microcomputers use stacks to handle data in an organized manner. A *stack* is a portion of the computer's memory that is used in a way similar to registers. They are last-in, first-out (LIFO) memories, which means that we can add items to the top or remove them from the top. To add an item to the stack is a PUSH operation; to remove a byte is a POP or PULL operation. A plate dispenser at a salad bar or cafeteria is similar to a stack in some microprocessors. When a plate is added or removed, all the plates move. A shift register is this type of memory or stack. An example of a LIFO stack is shown in Figure 5-3.

Most microcomputer stacks do not move. They are simply memory locations addressed by the stack pointer (SP), which is a register used only to keep track of the uppermost stack memory location. The microcomputer adds an element to the stack by placing the element in the memory location addressed by the SP and then incrementing the SP. The stack operation using an SP is shown in Figure 5-4.

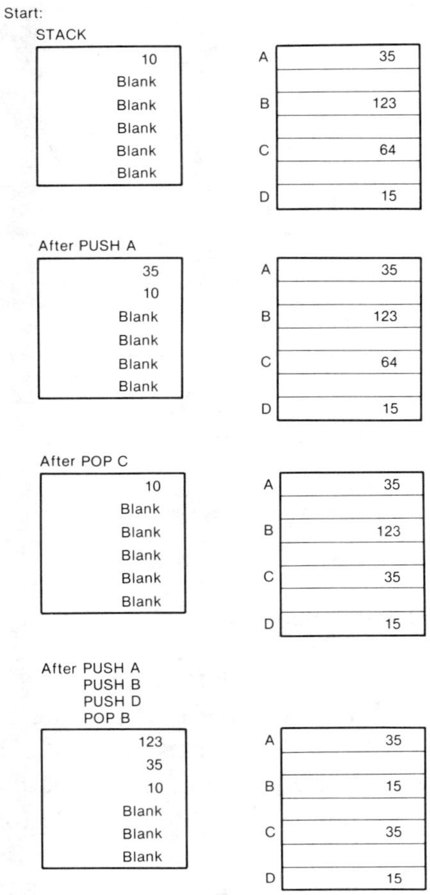


**Figure 5-3.** LIFO stack operations.

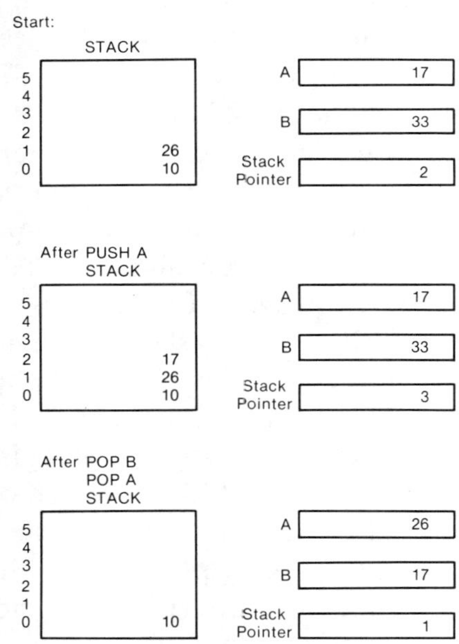


**Figure 5-4.** Stack operations using stack pointer.

To POP (remove) a byte from the stack, the SP is first decremented by 1 and the element is removed from the stack location being addressed by the SP. Recall that if we store data in a memory location or register, the data replace the previous contents and that previous data are lost if not saved prior to storing the new data. The major advantage of a stack is that we can add data to it without losing the data already stored there as long as the capacity of the stack is not exceeded. Also, data can be easily and quickly transferred to or from the stack because the address is always available in the SP and need not be a part of the instruction.

*Program Manipulation Instructions*—All of the instructions discussed until now have been "straight-line" instructions that execute one after the other in the order that they appear in the program. The real power of a microprocessor comes from its ability to execute a technique called *looping*. Looping allows a section of the program to run as often as needed. Every microprocessor has a group of instructions called *jump* and/or *branch* instructions that allow it to execute these program loops. These instructions allow the microprocessor to leave the main, straight-line path of the instruction sequence and branch or jump to a point in the program before or after its present location. Branching usually uses relative addressing. In the forward direction, the relative address is determined by subtracting the originating address from the destination address.

A backward branch is used when part of the program is to be repeated. The technique used for branching backward is similar to that used for branching forward except that the relative address is a negative number. A 0 in the sign bit will tell the microprocessor to branch forward, whereas a 1 in the sign bit tells it to branch backward.

The branch instructions may be conditional or unconditional. The unconditional branch instruction always results in a program branch when it is encountered in the program sequence. The conditional branch instruction causes a branch only if some specified condition is met.

A commonly used application is with the "branch if equal to 0" (BEQ), which causes a branch after a subtraction or a decrement if the 0 (Z) processor status flag is set to 1. If the result of the instruction (subtract or decrement) is not 0, then the BEQ instruction has no effect. Program flow can now take one of two paths, depending on the result of the subtraction. Conditional branch instructions give the microprocessor the capability of making decisions.

The jump (JMP) instruction is somewhat like the unconditional branch instruction. The difference is the method of addressing: the branch instruction uses relative addressing, whereas JMP can use either the indexed or the extended addressing mode. The jump instruction is often used to go to a subroutine. A *subroutine* is a group of instructions that performs a limited, usually short, though frequently required task. Subroutines are used so frequently that most microprocessors have several special capabilities that allow them to handle subroutines efficiently. Jumping to a subroutine is often referred to as *calling* a subroutine.

Every microprocessor has other simple program manipulation instructions. Examples are NO OP (no operation) and HALT. The NO OP does nothing except use an instruction cycle and increment the program counter. You may think that a NO OP instruction would never be used. However, a common application of this instruction is to replace program codes in the development of a program. If the programmer decides that a particular instruction is not needed, it may be replaced with NO OPs, which will not affect any of the registers or memory locations. After the program is completed and is working properly, the programmer can rewrite it without the NO OPs. The programmer can also use NO OPs to leave space in a program for possible insertion of other program codes during the development. Another useful application of the NO OP is to achieve a time delay, which is about two clock cycles. If a delay of more than several clock cycles is needed, the NO OP instruction is impractical. Another method should be used, such as time-delay looping. The HALT instruction causes the computer to enter a suspended state until an external signal is received.

*Status Management Instructions*—All of the categories of instructions discussed thus far have affected either data or the program. Status management instructions affect neither of these; instead they change the status conditions of the computer. They select memory pages, designate registers for a particular task, or override external controls to enable or disable interrupts. The enabling or disabling of interrupts is the most commonly encountered status management instruction. Software interrupts are of a higher priority than hardware (external) interrupts and are not usually masked out. The software interrupts are actually subroutine calls in most situations. Most microprocessors disable the interrupt system during timing loops, during RESET, and whenever the program cannot be properly resumed if interrupted. On the 8086 microprocessor, a 1-byte instruction is initiated by execution of the GO instruction. This instruction gives the programmer the ability to execute a segment of a program. The GO command is typically used to execute programs in the latter stages of debugging.

An 8086 program manipulation instruction is STEP (ST). This command provides the programmer or technician with the ability to execute one instruction at a time.

After this introduction to the types of instruction available on most microprocessors, let's look at how the assembly language mnemonics are translated into binary machine language codes.

## Assemblers

Earlier in this chapter, we discussed the advantages of assembly language over machine language. Remember that the microprocessor understands only machine language, a series of 1s and 0s. The basic purpose of an assembler is to translate the assembly language mnemonics into binary machine language codes. The computer can do this conversion if we supply it with a translation program.

Here is how this conversion is accomplished. The source program is compared with the translation program, which is known as the *assembler*. The assembler contains the mnemonics and the corresponding binary number for each mnemonic. When a mnemonic is loaded, the computer searches until it identifies that mnemonic. It then selects the corresponding binary numbers. The final result is a compilation of the source program written in machine language known as the object code. The mechanics of the process is shown in Figure 5-5. First, the assembler (translation program) is loaded into the computer and stored in memory. Next the source program written in assembly language is loaded and likewise stored in memory. Now the computer runs, translating the source program into machine language providing the object program. Note that this process does not execute the program but merely changes it into machine language. The exception to this is the "load-and-go" assembler, which places the object program somewhere in memory and starts executing it at some stated location.

The assembler can provide all or most of the following outputs:

- A listing of the assembly language program, together with the object code generated
- A list of assembly errors
- A symbol table with the meanings of all names used in the program
- A cross-reference table with a list of names and all instructions that use them
- A list of references to variables or subroutines defined outside the program
- A list of subroutines or macros used, and their lengths

Though most assemblers provide some features other than simply translating from source program to object program (such as labeling, assembly time arithmetic, comments, conditional assembly, and macros), most microprocessor designers and technicians generally use most of these features in their simplest form.

Most assemblers proceed by first processing the source code and reprocessing the outcome of the first step. The two steps are called *passes*, and assemblers that follow this procedure are called *two-pass assemblers*. The first pass consists primarily of noting the location assigned to each symbol. The second pass translates the op code and uses the information collected during the first pass to fill in the addresses in those instructions in which symbols were used to represent addresses.

The *field structure* of an assembler is the format of the locations on a program line: the label field, assembler op code field, address field, and comment field.

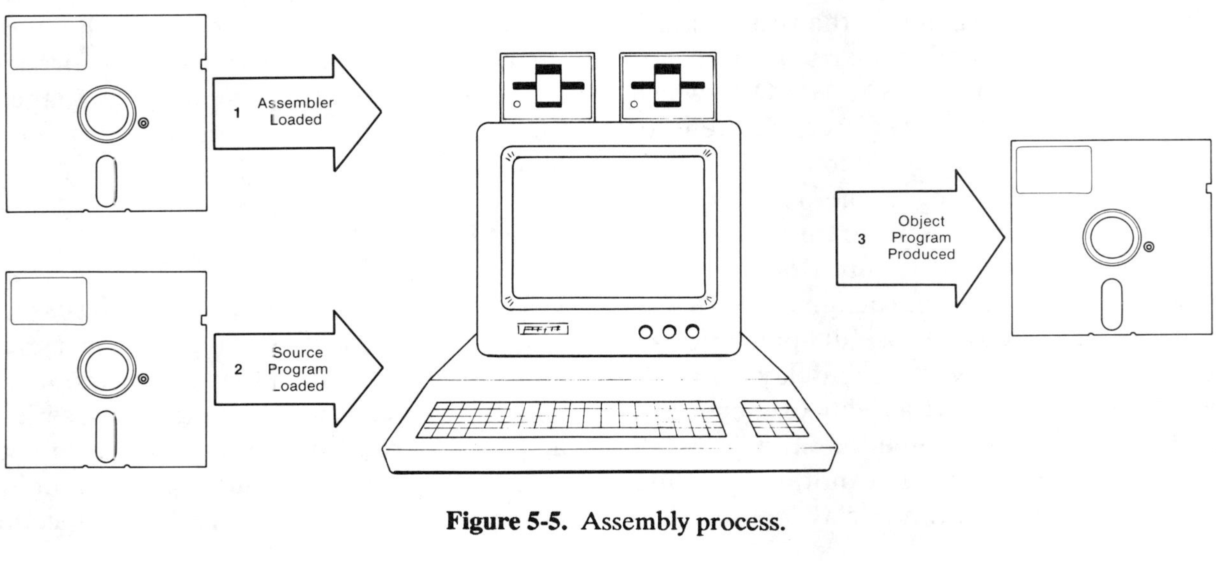


**Figure 5-5.** Assembly process.

The label and comment fields are optional in most cases, though some operations must have labels. A case where the label field is required is a *pseudo-operation*. The pseudo-operation is a directive to the assembler. It can assign programs and data to areas of memory, define symbols, and allocate space for variables. It can also determine the form of a program listing, generate fixed tables and data, and mark the end of a program. The specific names that assemblers use for these pseudo-operations vary greatly but a few common examples are

| | |
|---|---|
| **ORIGIN:** | allows us to start programs or data at a particular memory location |
| **EQUATE or DEFINE:** | defines names for later use in the programs (these definitions are usually placed at the beginning of the program so they can be easily found) |
| **RESERVE:** | sets aside RAM storage for variables |
| **END:** | marks the end of an assembly language program |

### Condition Codes

All of the registers, including the status register, in our hypothetical microprocessor were discussed in Chapter 3. However, the processor status register needs elaboration with regard to software and how it affects the operation of the microprocessor as certain instructions are executed. A couple of examples will give you the idea of how the flags are used in software.

The 8086 flag register was shown in Figure 3-5. Most instructions that process data and some that manipulate data affect the condition code flags. For example, the Z flag is set to 1 by the ALU if the last operation it performed produced a 0. If you subtract two numbers that are equal, the Z flag will be set to indicate that the result of the subtraction was 0. Also, if you AND two words together and the results contain no 1s, the flag will be set to indicate that the result was all 0s.

Another useful instruction that affects the 0 flag is the compare (CMP) instruction. The 0 flag will be set if the two operands compared are equal.

The DEC (decrement) instruction is another important instruction that affects the 0 flag. This instruction subtracts 1 from a number in a specified register or memory location. If, after decrementing, the contents of the register or memory location is 0, the Z flag will be set. For example, suppose that a series of actions needs to be repeated 12 times. To accomplish this, first load a register with the number 0CH (12 in hexadecimal) and execute the sequence of actions. Then decrement the register and look at the 0 flag to see if the register is down to 0. If the 0 flag is not set, the microprocessor is told to go back and execute the sequence of instructions again. This action can be accomplished with a conditional jump instruction.

Now, let's look at the carry flag with add, subtract, and compare instructions. This flag is also a multipurpose flag that the microprocessor can use in several diferent ways. It is used to indicate a carry out of the MSB when two 8-bit or 16-bit numbers are added. The carry flag is also used to keep track of a borrow when subtraction is performed on unsigned binary numbers. If a number is subtracted from a smaller number, the carry flag is made 0 to indicate that a borrow occurred. If the subtraction did not require a borrow, the carry flag is set to 1. The carry flag also takes part in various shift and rotate instructions. There are usually single-byte instructions available that allow the C flag, as well as other flags, to be set or cleared at any time.

The effects and instructions on status flags vary greatly from microprocessor to microprocessor. An important variation in how instructions affect status flags is whether they are affected by the groups of instructions—data transfer, program manipulation, and status management instructions. In some cases, we must save the status flag condition or check its condition immediately to avoid losing the particular needed status information. Though the effect of instructions on the status flags is peculiar to each computer, each seems to have its own advantages over the others, so none is clearly superior.

# Diagnostic Software

Diagnostic programs often come as software or firmware. Whatever the form, it uses the computer or microprocessor system to analyze itself at high speed to locate a malfunction. A software diagnostic program on diskette or tape is more detailed and allows the technician more options as to what to check and how to check it than does a program stored in ROM.

The basic principle of this testing technique is to compare an existing board, component, or system with what should be.

## Self-Diagnostics

When the self-diagnostic procedure is used, the microprocessor system, computer, or test instrument runs a set of tests to determine if it is itself properly operational. If a part of the self-diagnostic fails, then the program determines which part of the system is defective. An important portion of microprocessor system design and repair is the ability to test its memory. This involves determining whether a memory device and the associated memory selection logic are functioning properly.

An example of this memory testing is found in the IBM personal computer. IBM calls this the POST—power-on self-test. POST is found at the beginning of its operating system program near the reset location and performs a memory test each time the system is powered-up. It takes from 13 to 90 s to complete, depending on the amount of memory installed. It does a short test of the system board, memory, primary display, keyboard, diskette drive, fixed disk drive, and expansion unit. As it goes through the tests and the blocks of memory successfully pass, the memory blocks are counted on the screen. Then, when all of the system memory has passed this static test, a prompt is displayed along with a short beeping sound indicating that the self-test is complete. Whenever a POST failure occurs, and incorrect audio response, a blank screen, or an error message will occur. If an error indication is received during or at the completion of POST, note the error message and go to START on a troubleshooting tree provided in the manual. Follow through the steps indicated.

Other examples of self-diagnostic programs at power-up are in some microprocessor-based test equipment, such as digital storage oscilloscopes (DSOs) and logic analyzers. Typically, the DSO will run a self-test upon power-up to check that it is operational as well as calibrated. The logic analyzer's menu display reveals two important power-up conditions: the result of the mainframe power-up self-test and the configuration of the mainframe's installed data acquisition cards. Figure 5-6 shows a menu display of this type.

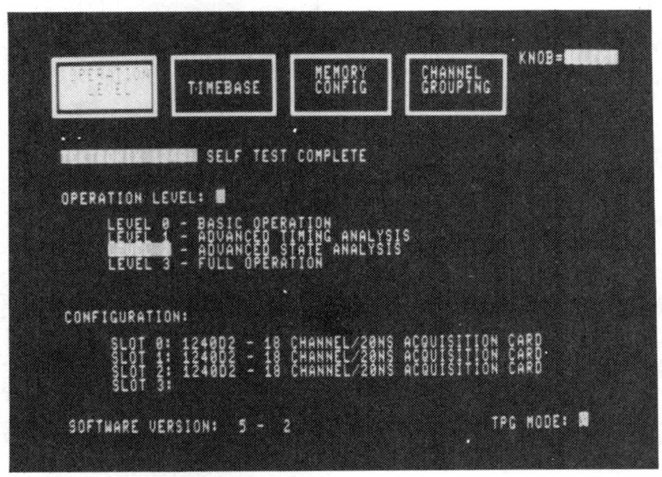


**Figure 5-6.** 1240 Logic analyzer power-up.

Whether an operational check or calibration check, the basic principle of a self-diagnostic is to execute a worst-case sequence and observe the results. Such a test will typically sequence through and exercise all of the instrument's or system's instructions in a predetermined order.

To test ROM, since it contains a known set of data, the individual bytes are either compared by adding them and checking the results against the stored correct value or by Exclusive-ORing the bytes together and again checking the results. The correct result, in either case, is stored in one of the memory locations of the ROM for this purpose. It is possible for this static self-test to miss a ROM failure, but it will find most of them.

The RAM self-test is about as simple as the ROM. However, the RAM does not have a known set of data at power-up. This known test condition must be provided by a self-diagnostic routine. The routine can first clear all of the bits in RAM, then check to see if they are all clear. If it passes that part of the test, the routine can set all the bits and then check to see if they are set. If RAM passes both tests, then the probability is excellent that it is free of defective cells, especially if the cell is totally burned out. A marginally defective cell could slip past the static test. For example, some marginal defects of a cell might be that it forgets, after a few microseconds or milliseconds, thus losing the data. Or an electrical noise impulse generated by reading or writing to a nearby cell or other part of memory can change the contents of the memory cell. A dynamic memory test might expose this type of malfunction. A test of this type consists of storing data into a group of memory locations, then attempting to disrupt this data by writing to other locations in the memory.

## Monitor Systems and Diagnostic ROMs

In the evolution of microprocessor-based equipment, diagnostic programs and monitor systems stored in ROM have become more complex, powerful, and more commonly encountered. Here we describe some features and capabilities of some common to the 6800 microprocessor.

A monitor system should at least include the ability to

- Read the status flags
- Read the register contents
- Read the memory contents at any logic analyzer
- Single step through a program
- Set the starting address into the microprocessor
- Run the program at full speed

The unique feature of self-diagnostic programs is that they use the microprocessor to test its own system. A special diagnostic program will fetch dynamically the register information and display it after the system runs each instruction of the user program or the program under development. There are several ROM-based monitor programs available, though one of the simplest is a ROM IC, part number MCM-6830L8, readily available from distributors of Motorola ICs. This monitor program is called MIKBUG.

Other, more modern, powerful, and complex monitor ROMs are available, such as MINIBUG III. This monitor routine has simple breakpoints, which are described later in this chapter.

## High-Level Diagnostic Programs

High-level diagnostic programs are used to locate a problem in a computer or computer-based system. This type of program is more comprehensive and complex than the ROM-based diagnostic programs and allows the troubleshooting technician many test possibilities. The first requisite to use this valuable troubleshooting tool is that the computer system must be operational enough to load the diagnostic program from a floppy disk drive. Whenever first learning to use a diagnostic program, learn to run the routine in a system that is known to be in good operating condition. Become familiar with the various prompts and options available as the program runs the various tests.

A typical high-level diagnostic program diskette is the IBM Hardware Maintenance and Service. This advanced program helps the technician isolate and repair any failure of a field replaceable unit (FRU) in the IBM computer.

The first step is to load the advanced diagnostic software from the diskette. Menu 1 shown in Figure 5-7 will appear on the display. The 0 option is usually selected by a technician. Options 1 and 2 are similar to DOS programs for formatting a diskette and for copying one. Option 3 is of negligible value at this point, and option 9 is the way back to the operating system. If a 0 is selected (RUN DIAGNOSTIC ROUTINE), then the system checkout procedure begins and Menu 2, shown in Figure 5-8, is displayed. A Y response goes to system checkout menu 3 (Figure 5-9). An N allows the technician to add or delete items from the installed devices list. It requires a Y to continue. A 0 or 1 goes to menu 4 (Figure 5-10) for selection of options to be tested by entering corresponding numbers. To select all menu 4 options, press ENTER. If action 2 is selected from menu 3, the display goes to menu 5 (Figure 5-11), the LOG UTILITIES feature. This feature allows use of the error log and time of day function. Selecting 9 from menu 1 exits the diagnostic program and loads the alternative program from the diskette in drive A.

```
The IBM Personal Compute^
ADVANCED DIAGNOSTICS
Version 2.02 (C) Copyright IBM Corp 1981, 1983

SELECT AN OPTION

0 - RUN DIAGNOSTIC ROUTINES
1 - FORMAT DISKETTE
2 - COPY DISKETTE
3 - PREPARE SYSTEM FOR RELOCATION
9 - EXIT TO SYSTEM DISKETTE

ENTER THE ACTION DESIRED
? _
```

**Figure 5-7.** Menu 1.

Additional menus go from menu 4 through the specific options selected for testing with a step-by-step procedure. All are menu-driven. The only additional test equipment needed to go through the complete trouble diagnostics is a screwdriver, a VOM, and a tweezer-type module puller for removal of 16K and 64K memory modules.

```
THE INSTALLED DEVICES ARE

 1 - S SYSTEM BOARD
 2 - S 512KB MEMORY
 3 - S KEYBOARD
 4 - S MONOCHROME & PRINTER ADAPTER
 6 - S 1 DISKETTE DRIVE(S) AND ADAPTER
 9 - S PRINTER ADAPTER
11 - S ASYNC COMMUNICATIONS ADAPTER
12 - S ALT ASYNC COMMUNICATIONS ADPT
17 - S 1 FIXED DISK DRIVE(S) & ADAPTER

IS THE LIST CORRECT (Y/N) ? _
```

**Figure 5-8.** Menu 2.

```
SYSTEM CHECKOUT

0 - RUN TESTS ONE TIME
1 - RUN TESTS MULTIPLE TIMES
2 - LOG UTILITIES
9 - EXIT DIAGNOSTIC ROUTINES

ENTER THE ACTION DESIRED
? _
```

**Figure 5-9.** Menu 3.

```
 1 - S SYSTEM BOARD
 2 - S 512KB MEMORY
 3 - S KEYBOARD
 4 - S MONOCHROME & PRINTER ADAPTER
 6 - S 1 DISKETTE DRIVE(S) AND ADAPTER
 9 - S PRINTER ADAPTER
11 - S ASYNC COMMUNICATIONS ADAPTER
12 - S ALT ASYNC COMMUNICATIONS ADPT
17 - S 1 FIXED DISK DRIVE(S) & ADAPTER

ENTER THE NUMBER(S) OF OPTIONS TO TEST
OR PRESS ENTER TO SELECT ALL OPTIONS
? _
```

**Figure 5-10.** Menu 4.

```
LOG UTILITIES

0 - START ERROR LOG
1 - STOP ERROR LOG
2 - LIST LOG
3 - SET TIME OF DAY
4 - DISPLAY TIME OF DAY
9 - RETURN FROM UTILITIES

ENTER THE ACTION DESIRED
? 2
DISKETTE OR CASSETTE LOG (D C)? _
```

**Figure 5-11.** Menu 5.

## A Floppy Disk Drive Diagnostic Program

A valuable troubleshooting aid to the technician who regularly repairs floppy diskette drives is a package called the *interrogator*. The two diskettes of the package are a menu-driven software program and a digital diagnostic diskette (DDD). Together these can be used to evaluate the flexible disk drives in the IBM PC, XT, or PC portable.

The two diskettes in this package are intended to be used by the technician to check a drive's head alignment, spindle speed, centering, and its ability to read and write data. Using this diagnostic system, the technician can make all of these tests without disassembling the computer or using any other test equipment.

To make the testing most effective, the publisher of this diagnostic program recommends its use in the following manner. First, use the auto sequence tests, which quickly check the drive's alignment using the DDD disk. This test links together a spindle speed test, a centering test, a radial alignment test, an azimuth alignment test, and a hysteresis test. After these, the R/W tests should be used, which continuously read and write randomly chosen sectors. Any errors found during this test will help isolate intermittent mechanical and electrical problems. If the diskette drive under test is suspect of having a R/W problem, which could destroy the DDD, the R/W tests should be run first. Other tests are available with the interrogator to help isolate the problems found during these first tests.

Many of the methods presented here point to a trend in troubleshooting digital and microprocessor systems that includes an increased dependency on self-testing in both software and hardware. If the system does not contain the necessary self-diagnostic capabilities, there will probably be auxiliary computer-driven test equipment to perform the diagnosis and testing on the system until the fault is located. However, the technician will still need the analyzing ability and digital testing skills to make the actual repair and get the equipment back in service. The simpler the system, the more limited the automatic diagnostic features it will have so the troubleshooter must possess more skill.

# 6

# Using Some Basic Digital Troubleshooting Instruments

The basic test instruments used to troubleshoot digital equipment are essentially the same as for other solid-state units. Much of the service work can be handled with a good DVM and an oscilloscope. These are briefly reviewed here, and items made specifically for digital troubleshooting are discussed, including a logic probe, a current pulser, a current tracer, a logic clip, and a comparator. In digital communications, a bit error rate tester (BERT) and a breakout box are essential to troubleshooting. These instruments are also discussed briefly.

## The Digital Multimeter (DMM)

In digital circuits, the same basic electrical laws are at work as in analog circuits. To know something about the electrical properties and parameters of a circuit requires making measurements. The static measurements of voltage, current, or opposition can be made with one of two basic multimeter types: analog (a VOM) or digital (a DMM). The main part of a VOM is the meter movement (usually d'Arsonval), which has a needle that deflects along a scale and indicates the quantity of the parameter by its position or amount of deflection. Many tests on digital equipment may be made with a VOM, but the DMM is much more practical for most digital troubleshooting. The technician should retire the VOM for working on digital equipment. "Yes, but . . . ," you say? Alright, keep it around. Every digital circuit has a power supply, for example, and the VOM can still do some tricks that give it the edge over the DMM. We will consider some of these tricks in a later chapter. In this chapter, we will look at applications and characteristics of the DMM.

In Chapter 1, we considered logic voltage levels that represented a valid high and a valid low. These values required an accuracy and resolution that would be difficult to achieve on a VOM, though they are easily achieved even on an inexpensive DMM like that in Figure 6-1.

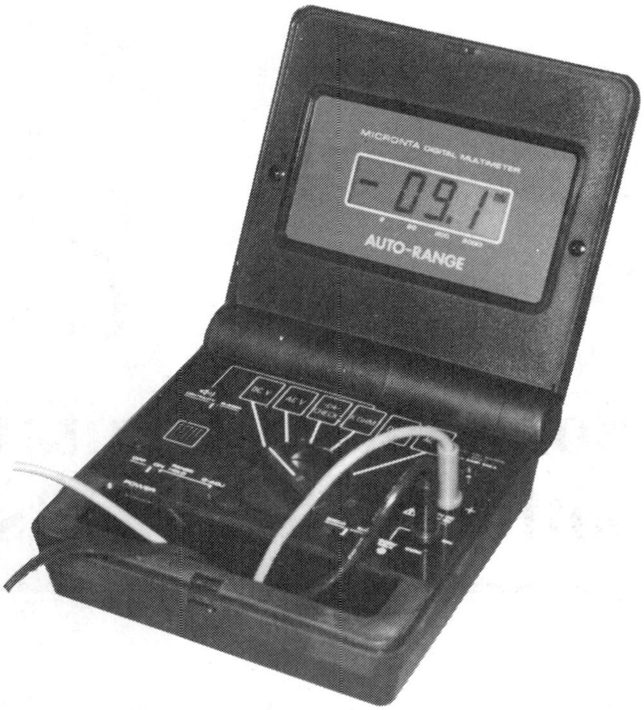


**Figure 6-1.** Digital multimeter.

The change in the voltage level as one gate or one input of a fan-in goes high or low may be displayed on a DVM with $2\frac{1}{2}$ digits of resolution. The accuracy on the DMM remains the same throughout the range (+ or − one digit). However, on the VOM it depends on the range and from what part of the scale the reading is taken. Another advantage or characteristic of the DMM is that there is no parallax error (error due to interpolation between scale markings). The DMM has a high internal resistance that remains constant on all ranges, so the loading effect, though constant, is minimal.

The DMM in Figure 6-1 has a liquid crystal display (LCD) that uses very little power from the batteries but must have ambient light to be visible. Some DMMs use light-emitting diodes (LEDs), which use much more power but are much brighter and not easily washed out in bright sunlight.

In addition to being used for measuring for valid logic levels, the DMM is excellent for chasing down shorts and opens on a printed circuit board (PCB). Remember that a short between two points means zero voltage and an open means zero current. A short between two runs on a PCB will produce a 0-V reading between the runs unless the DMM has a microvolt scale, which is unlikely. The DMM can then be used to monitor the very low resistance between the runs *with the power off* in the circuit while the short is found and cleared. An open PCB run can likewise be traced out by starting with the DMM probes together at one IC pin on the suspected run and then moving the probe down the run. An open, even a hairline crack, in the board can be found by this method quite easily. Tracing a short or an open is shown in Figure 6-2.

**Figure 6-2.** Tracing a short or an open.

# Oscilloscope

A good technician or service engineer tackling a tough troubleshooting problem will probably grab an oscilloscope. To a competent experienced user, the oscilloscope is the most valuable general-purpose test tool. Figure 6-3 shows an excellent example of an oscilloscope for digital troubleshooting.

The oscilloscope draws a graph of how much (amplitude) and how fast; that is, it plots on its screen voltage per division (vertically) as a function of time per division (horizontally). In

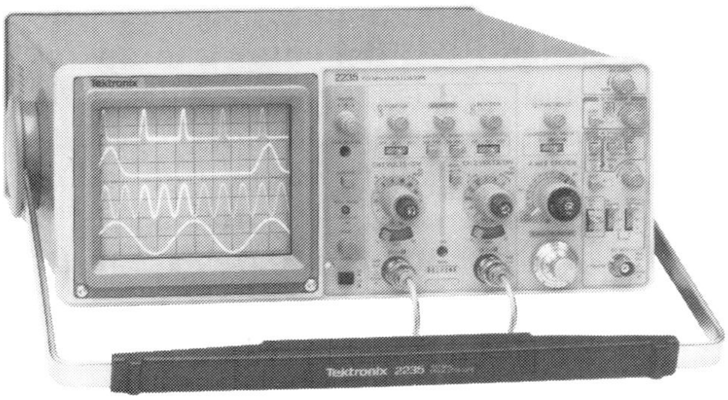

**Figure 6-3.** The Tektronix 2235 oscilloscope. *(Courtesy Tektronix, Inc.)*

combinational logic, we are often interested only in measuring the voltage levels, so we ignore the time axis. We first discuss some of the scope's vertical controls, along with how to make good measurement.

To get a trace on the screen, you must trigger the scope or set it to automatic (Auto) triggering and increase the intensity (Int) to a comfortable viewing level. The scope's position (Pos) control must be set for an on-screen trace.

Scopes are more useful if they have more than one vertical display channel. The modes of display are typically channel 1 only, channel 2 only, and both [in either alternate (Alt), chopped (Chop), or add (Add)]. The Alt and Chop modes are provided to allow viewing of two signals at any sweep speed. The Add allows the two to be algebraically added together and displayed as one waveform. One of the channels may be inverted in polarity, and then Add will actually present the difference between channel 1 and channel 2. Figure 6-4 shows the detail of the vertical mode selection of the scope of Figure 6-3. It has two vertical channels.

To draw a graph, the scope needs horizontal and vertical data. The horizontal sweep system of the scope supplies the second dimension by providing the deflection voltages to move the electron beam horizontally. This horizontal dimension data is provided by the scope's time base or sweep generator. This sweep

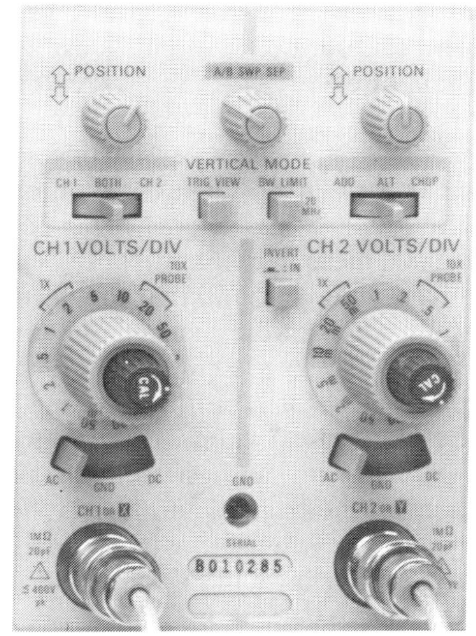


**Figure 6-4.** The Tektronix 2235 oscilloscope—detail of vertical mode selection. *(Courtesy Tektronix, Inc.)*

generator produces a *ramp*, sometimes referred to as a *sawtooth*, waveform to move the beam horizontally. This ramp voltage is shown in detail in Figure 6-5.

Notice the three parts of the sweep waveform: the rising portion called the ramp (trace), the retrace, and the holdoff (the time between the ramps including the trace time). The sweep rate of the beam from left to right is determined by the slope of the ramp. This is controlled by the Sec/Div switch. The return of the beam to the left side of the screen takes place during the

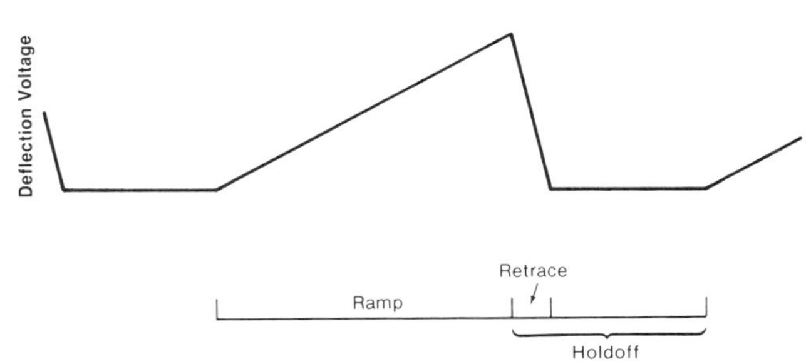


**Figure 6-5.** Ramp voltage used for horizontal sweep.

retrace. The holdoff period provides additional time beyond retrace to ensure a stable display.

The time-base generator allows you to choose the calibration time units. Just as having more than one vertical channel extends the scope's versatility greatly, a second time base for the horizontal system allows extra convenience in some measurements. However, sometimes a second time base, called *delayed sweep* is the only way to make a measurement.

So far, we have found that the vertical and horizontal systems of the scope supply the information to the display system (CRT) for drawing the graph or waveform on the screen. The only thing missing is the "when." When should the circuits start drawing, and when should they not? The "when" is the trigger, and it tells the sweep circuit precisely when to start drawing the waveform. This is shown in Figure 6-6.

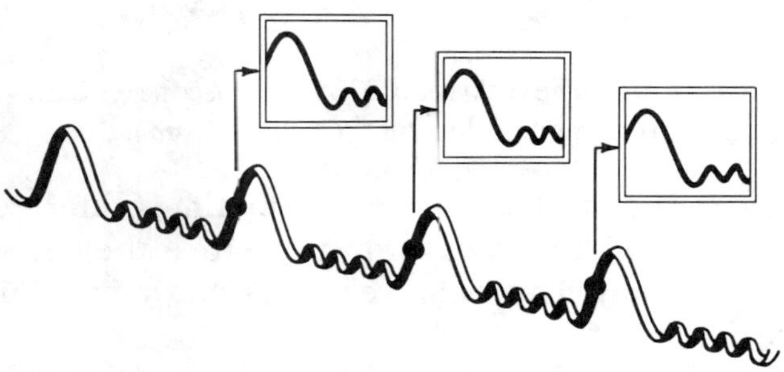

**Figure 6-6.** Detail of oscilloscope trigger.

The "displayed" part of the signal is the part that occurs during the ramp part of the sweep waveform, as we saw in Figure 6-5. If each sweep started at a different place on the signal, it would produce a confused jumble on the screen that would be meaningless. To ensure that the sweep starts at the right time, you must make the right trigger system control settings. Most scopes have five trigger control settings: Mode, Source, Coupling, Slope, and Level. See Figure 6-7.

Oscilloscopes may have several trigger modes, some of which are for special applications, and may be named single-sweep, TV (Field and/or Line), and so on. However, the most useful will be the normal and automatic modes. The normal trigger mode (Norm on the Mode switch) will be the most useful because it can handle a wider range of trigger signals than any other trigger mode. It gives the widest range of triggering signals and a steady signal display at any repetition rate. The only disadvantage in using Norm triggering is that when there is no triggering signal applied to the trigger circuit, the screen will be blank—no trace at all. The automatic mode (labeled Auto on the front panel) takes care of this disadvantage. In Auto mode, sometimes referred to as "bright baseline" mode, a trigger starts a sweep. The sweep ends, and the holdoff period expires. At that time, a timer begins to run. If a trigger signal is applied to the trigger circuit before the timer runs out, a trigger is generated anyway, causing a bright baseline or reference trace to appear on screen. This occurs even when there is no triggering signal available (i.e., no signal applied to the vertical channel).

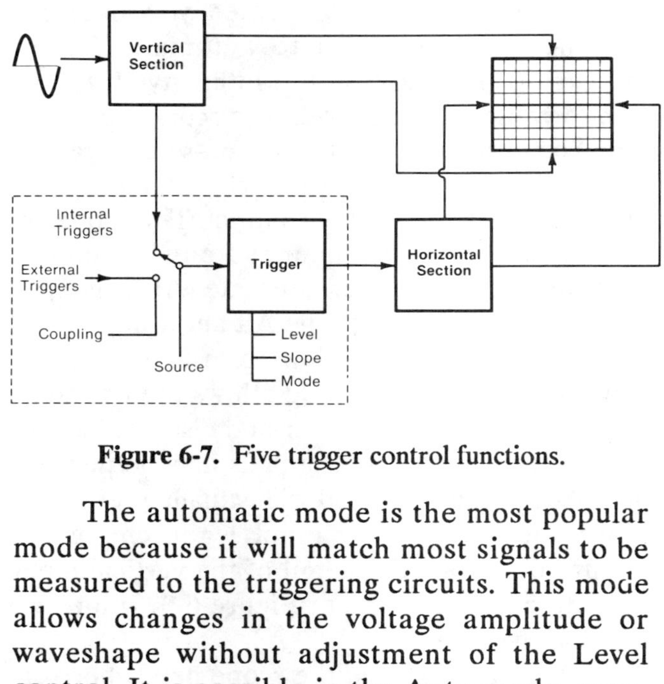


**Figure 6-7.** Five trigger control functions.

The automatic mode is the most popular mode because it will match most signals to be measured to the triggering circuits. This mode allows changes in the voltage amplitude or waveshape without adjustment of the Level control. It is possible in the Auto mode, especially on low-frequency signals with a repetition rate mismatched to the Auto mode timer, to have a display that is not steady. The Norm mode gives a steady trace on screen at any "rep" rate.

The necessity for holdoff time was seen in Figure 6-5. Sometimes, the normal holdoff time is not long enough to ensure getting a stable display. This possibility can exist when the triggering signal is a complex waveform with many possible trigger points on it. The digital pulse train shown in Figure 6-8 is a good example. Though the waveform is repetitive, a simple trigger with a fixed holdoff time might reap a series of patterns on the screen instead of the same pattern each time.

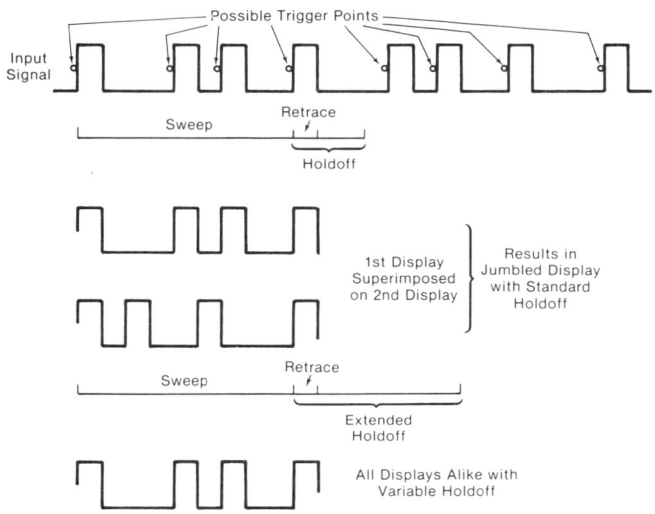


**Figure 6-8.** A jumbled display because of fixed holdoff.

These multiple patterns would result in a jumbled display. A longer holdoff time is needed. The Variable Holdoff Control allows the scope to ignore some potential trigger points. This control is actually part of the horizontal system of the scope because it adjusts the holdoff time of the sweep generator, though its function interacts with the trigger controls. Without the Variable Holdoff Control, a stable trace can be obtained only by turning the Variable Sec/Div Control out of Calibrate and adjusting it for a possible stable display. With the time/division now uncalibrated, a meaningful time measurement cannot be made. The Variable Holdoff Control is quite valuable in changing the holdoff time and keeping the trace (sweep) calibrated. The trigger point will appear on the same pulse in each repetition of the input signal, thus ensuring a stable waveform display.

With the waveform display stable, time and voltage measurements can now be made. For the greatest accuracy, use the center graticule lines, with the signal covering as large an area on the screen as possible. The two most basic measurements are amplitude and time. Almost every other measurement, such as frequency or duty cycle, is based on one of these two fundamental techniques.

To make an amplitude measurement, connect the probe to the input BNC jack of one of the vertical channels and select that channel with the Vert Mode switch. Make sure the Variable Volts/Div control is in the "detent" or CAL position to make a measurement. This variable sensitivity control is useful when making quick amplitude comparison of a series of signals or setting the relative amplitude of a signal to a convenient height to make time measurements. Remember to put the vertical variable controls back to CAL before making any amplitude measurement.

The probe must be compensated if it is an attenuator probe (×10, for example). The probe should be compensated because it is going to be used when making the measurement. Compensate it with the accessory tip to be used. Do not compensate the probe in one vertical channel and then use it in another. Failing to properly compensate the probe is the most common mistake in making oscilloscope amplitude measurements. The procedure is described in Figure 6-9.

(A) Input pulse. (B) Undercompensated probe. (C) Overcompensated probe.

**Figure 6-9.** Compensating an oscilloscope probe.

Another common error is assuming a proper ground in making the measurement. Make sure the scope ground is the same as the circuit ground—do not just assume that it is. Check the circuit ground by touching the probe to the point assumed to be ground *before* making a hard ground—that is, before connecting the oscilloscope ground strap of the probe. The vertical coupling should be DC, and the vertical Volts/Div should be set to the most sensitive position. The trace should not move if the circuit ground is the same as the scope's earth ground. If the trace does move, check to see if the right point is being probed for circuit ground and that the proper power cord and plug is being used on the scope. If a scope cart is used, sometimes the proper power ground will have been defeated. This produces not only measurement error but a potentially dangerous situation.

In oscilloscope measurements, amplitude usually means peak-to-peak amplitude. Connect the probe to the signal to be measured. Since most digital signals are pulses, assume that it looks something like the signal in Figure 6-10. Be sure that the coupling switch is set to DC if the total amplitude of the pulse, not just the AC component, is needed.

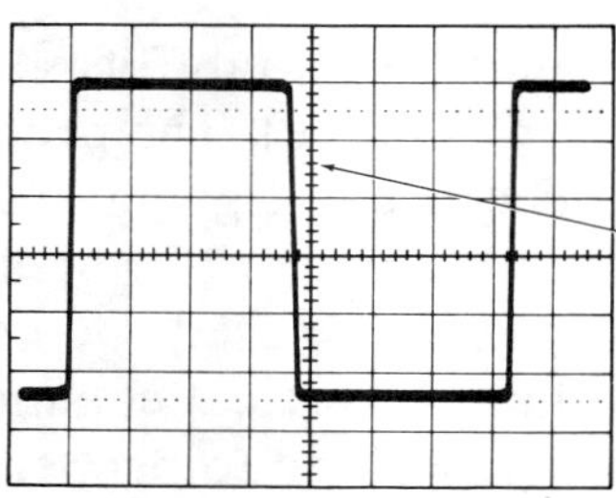


**Figure 6-10.** A pulse waveform–vertical voltage measurement.

The 0-V reference is set with the Vertical Position control and the coupling switch at the GND position. It may be convenient to set the reference in the center vertically if the signal is going positive and negative from 0 V. If the signal is always positive, it may be more convenient to set 0 V (GND) at the bottom of the graticule. If the signal is always negative, then the top of the graticule may be the best place for 0 V.

Set the Volts/Div switch so that the signal is something less than the full screen vertically. The signal should be as large as possible for the highest resolution. Set the trigger Source to INT and to the vertical channel to which the signal is connected. The trigger Mode switch can be set to Auto, and the trigger Level control can be adjusted for a stable trace. Turn the Sec/Div switch until approximately two cycles of the waveform are on the screen.

Use the Horizontal Position control to move the signal until the portion to be measured is nearest the center vertical graticule line. Count the major and minor divisions (Div) down the center vertical graticule line and multiply by the Volts/Div setting to make the measurement. For example, in Figure 6-10, the peak-to-peak signal amplitude is 5.5 Div × 0.1 V/Div, which is 0.55 V.

Time measurements may also be made from this setup for amplitude measurements. To make a period measurement, for example, line up one rising edge of the pulse signal with a vertical graticule line near the left side of the screen with the Horizontal Position control. Set the Sec/Div control so that the next rising edge of the pulse crosses the center horizontal graticule as far to the right side of the screen as possible. Count major and minor divisions from left to right, as shown in Figure 6-11. Multiply the number of divisions by the Sec/Div setting. For example, in the figure, 7.2 div × 0.2 ms/Div = 1.44 ms.

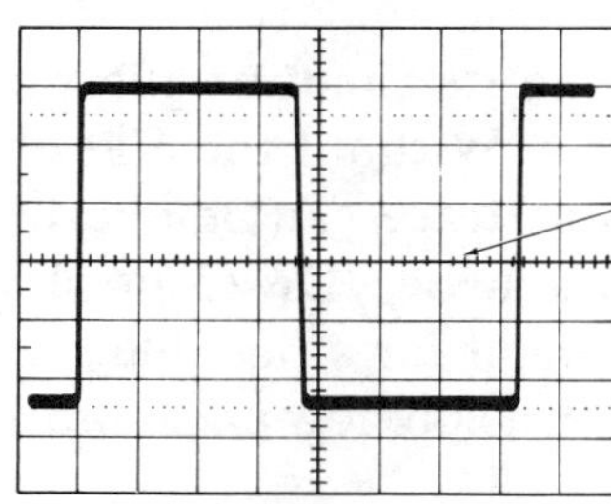


**Figure 6-11.** A pulse waveform–horizontal time measurement.

In the troubleshooting of digital equipment, pulse measurements are important because the circuit can intentionally or unintentionally modify or degrade a digital pulse. The most important signal parameters of a pulse are shown in Figure 6-12. *Amplitude* is the amount of displacement from a reference without regard to direction. In oscilloscope measurements, amplitude usually means peak-to-peak amplitude. Peak and peak-to-peak for the pulse are shown from the dotted-line equilibrium at a particular point in time. *Preshoot* is a change of amplitude in the opposite direction that precedes the pulse. *Overshooting* and *rounding* are changes that occur after the initial transition. *Ringing* is a set of amplitude changes, usually a damped sinusoid, that follows overshooting. All of these are expressed as percentages of amplitude. *Settling time* expresses how long it takes the pulse to reach its maximum amplitude. *Droop* is a decrease in the maximum amplitude with time. Nonlinearity is any variation from a straight line drawn through the 10% and 90% points of a transition.

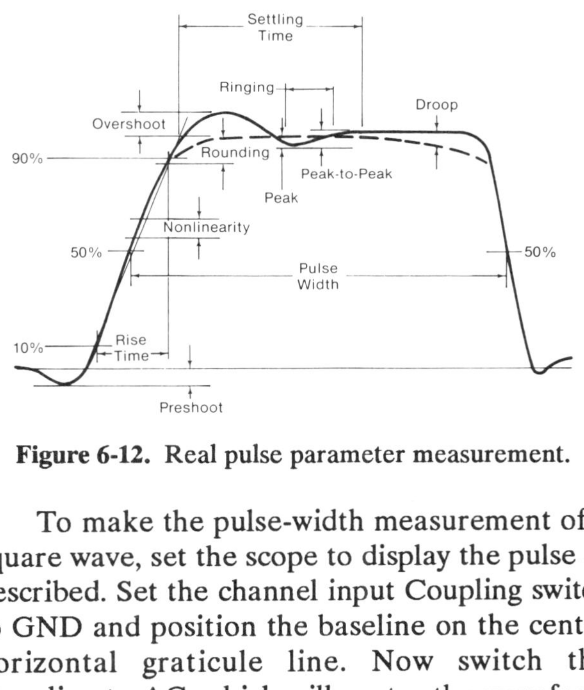


**Figure 6-12.** Real pulse parameter measurement.

To make the pulse-width measurement of a square wave, set the scope to display the pulse as described. Set the channel input Coupling switch to GND and position the baseline on the center horizontal graticule line. Now switch the Coupling to AC, which will center the waveform vertically if it has a 50% duty cycle (i.e., positive and negative alternations equal in time. The pulse-width measurement is made at the 50% amplitude point of the waveform. Count major and minor divisions across the center horizontal line and multiply by the Sec/Div for the pulse width.

If the pulse is not a square wave (50% duty cycle), use the vertical position control to center it vertically. The Variable Volts/Div may be turned out of detent (uncalibrated) to adjust the amplitude for a convenient height, using the vertical major divisions to make sure that the same amount of the pulse is above and below center. Now make the measurement of pulse width on the center horizontal graticule line as before.

To obtain the DC offset of the waveform, switch the coupling to DC. Any vertical shift represents the DC offset or DC component of the waveform. Count the divisions of shift on the center vertical graticule marker and multiply by the Volts/Div setting to find the DC offset.

# Delayed Sweep Measurements

Delayed sweep is a technique that adds a precise amount of time between the trigger point and the beginning of the scope's sweep. Often delayed sweep is used as a convenient way to make a measurement. However, sometimes it is the only way a measurement problem can be solved.

The delayed sweep feature utilized most often is the intensified sweep used as a positionable magnifier. Trigger the scope normally. Then switch to the intensified horizontal mode. Run the delay time, and therefore the intensified zone, out to the part of the signal waveform in which you are interested. Then switch to the delayed mode and increase the sweep speed of the delayed sweep to magnify the selected waveform portion so that it can be examined in detail.

Dual-time-base scopes normally have delayed sweep as a standard feature. They offer all the measurement capabilities of a single-time-base scope plus

- Convenient comparison of signals at two different sweep speeds
- Jitter-free triggering of delayed sweeps
- Timing measurement accuracy of near 1%

To keep these two time-base sweeps separate while discussing them, we refer to the first one as the delaying (A) time base and to the second as the delayed (B) time base. The length of time between the start of the A sweep and the start of the B sweep is called the *delay time*. Each of the three measurement capabilities mentioned will be discussed briefly.

### Measurements at Two Sweep Speeds

Looking at a signal with two different sweep speeds makes complicated timing measurements easy. The A sweep is a large slice of time on the signal to be examined. The intensified zone shows where the B sweep is positioned. The faster B sweep speed magnifies the smaller portion of the signal in great detail. Using sweeps A and B and dual trace from vertical channels 1 and 2 allows four traces to be displayed.

The scope of Figure 6-3 has an additional position control labeled Alt Swp Sep for alternate sweep separation. With this control and the two vertical channel Position controls, all four traces can be positioned on the screen without confusion.

### Delayed Sweep Triggering

Time jitter can prevent an accurate measurement of a signal that is not perfectly periodic. This problem can be solved with the separate trigger available for the time base B. In our example application, the B sweep ran immediately after the delay time. On the scope of Figure 6-3, the Runs After Dly mode is selected by turning the B Trigger Level control to its full clockwise position. At any other position, it functions to set the trigger level for the B trigger. Along with the B Trigger Slope Control, a stable B sweep can be triggered even when the A sweep has jitter.

### Increased Timing Accuracy

Most dual-time-base oscilloscopes have a measuring indicator dial that positions the intensified zone (B sweep) on the A sweep display. This indicator is usually a 10-turn helipot that has a window on top with numbers corresponding to the horizontal major divisions of the scope screen. The numbers around the circumference divide the major divisions into hundreds. Some scopes today have a digital readout of the delay time. This control on the scope of Figure 6-3 is labeled B Time Delay Position. To make a 1.5% accurate measurement on this and many similar scopes,

1. Use the B Runs After Dly mode.
2. Place the intensified zone where the timing measurement begins (or use the B sweep waveform and the center vertical graticule line) and note the B Delay Time Position dial setting.
3. Dial back to where the measurement ends and note the reading there.
4. Subtract the first reading from the second reading and multiply by the A sweep Sec/Div setting.

### More Technically Sophisticated Oscilloscopes

The scope in Figure 6-3 has many standard features that enable it to make almost every measurement needed for digital troubleshooting. The oscilloscope market is very competitive. New scopes with more features at lower cost, higher bandwidths, and faster measurement capabilities appear continuously. The options available include cursors and automatic calculation of waveform parameters. An example is shown in Figure 6-13.

Another complete family of oscilloscopes is the digital storage oscilloscope group.

## Digital Storage Oscilloscopes

Storage oscilloscopes are used when signals are too slow or too infrequent for conventional

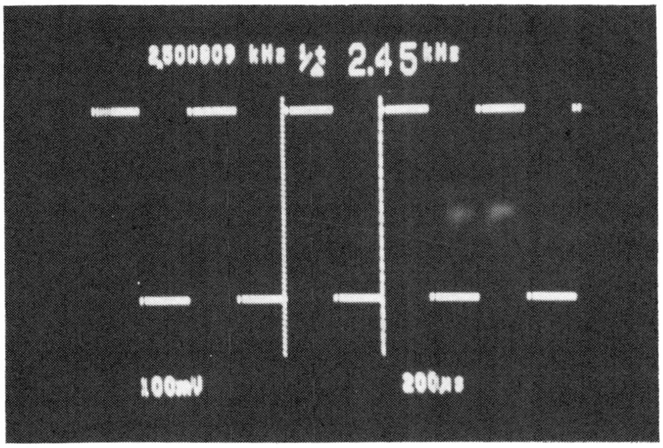


**Figure 6-13.** A waveform showing frequency in hertz displayed when cursors are used.

oscilloscopes to capture and when the technician needs to compare events that occur at different times instead of simultaneously. Storage instruments include CRT storage osilloscopes and digital storage oscilloscopes (DSOs). In CRT storage scopes, the electron beam writes on a target within the CRT to capture the waveform. The DSO quantizes (digitizes) the waveform and stores it in a digital memory. Having quantitized waveforms in a digital memory gives you measurement and analysis capabilities not possible otherwise.

Although laboratory DSOs have been available for several years, they have not been considered as service tools until now. The reasons for this change in view is because price in portable DSOs has dropped significantly due to maturing technology. Also, the range of built-in features simplifies many measurement and troubleshooting procedures. We will describe the features particularly important to digital troubleshooting.

Waveform digitizing and storage is a key feature for simplifying waveform measurement and comparison tasks. To help with this job, cursors are available for on-screen measurement.

Peak detection is one of the most important features for digital troubleshooting. Narrow glitches, a common cause of digital circuit malfunctions, are often missed with standard oscilloscopes but are easily captured with DSO peak detection. Because peak detection can find troubleshooting problems that would otherwise go undetected, this feature alone can justify investment in a DSO as a troubleshooting tool. The ability to capture and display glitches as short as 2 ns at any sweep speed means that you do not miss unexpected causes of circuit malfunctions. The oscilloscope shown in Figure 6-14 has this capability.

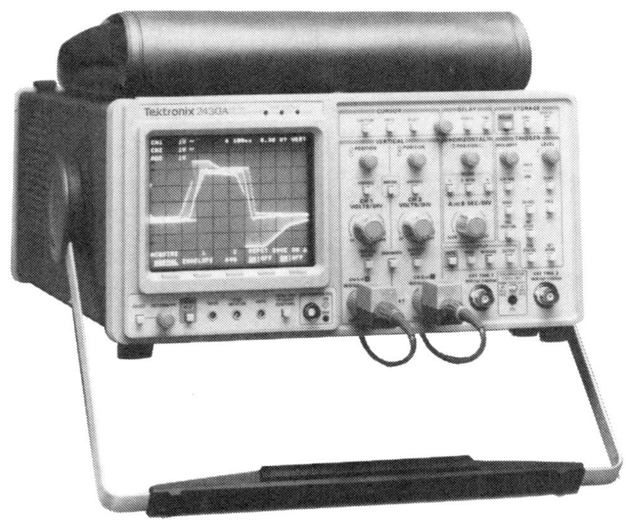


**Figure 6-14.** The Tektronix 2430A digital storage oscilloscope. *(Courtesy Tektronix, Inc.)*

Digital storage oscilloscopes allow pretrigger viewing; that is, you can look at the waveform both before and after the trigger event. Also, the DSO trigger can stop as well as start an event—a feature known as *babysitting*—which is very useful for intermittent troubles.

Another feature of DSOs not available in conventional oscilloscopes is the ability to include digital processing, such as averaging samples of the input signal to reduce the effects of random noise. This feature is shown in Figure 6-15. Calculations can be performed on the waveform parameters and the signal data and then be output over RS-232 or GPIB standard interfaces. This process is called *computer-based automation*. Computer-based automation could well be one of the most useful and cost-effective troubleshooting features of DSOs. A simple troubleshooting application is an extension of storing standard waveforms for comparison at a later time. Instead of powering-up a standard or working device and acquiring its characteristic waveform, the waveform can be stored on a computer disk. A whole library of standard and reference waveforms (previous trouble cases, for example) can be stored and then downloaded to the DSO's memory when needed.

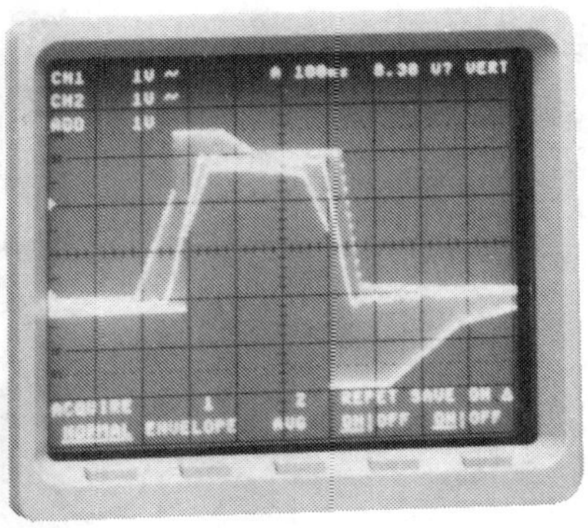


**Figure 6-15.** Waveform showing some DSO features.

From the very few DSO applications given here, you can begin to get a feeling for the tremendous range and depth of capabilities available from the use of a DSO in troubleshooting.

# The Logic Probe

The logic probe is a functional measurement tool for troubleshooting digital circuits. It has become as commonly used by digital troubleshooters as the screwdriver is by auto mechanics. In fact, it is becoming popular with the modern automobile mechanic because of the digital circuits in the computer-controlled car (CCC). The logic probe detects logic levels at any circuit point and displays the results (high, low, pulses) on LEDs. The logic probe in Figure 6-16 provides the necessary information at points in a digital circuit to locate most problems. It costs less than $20.

The probe has variable threshold levels because CMOS and other families sometimes

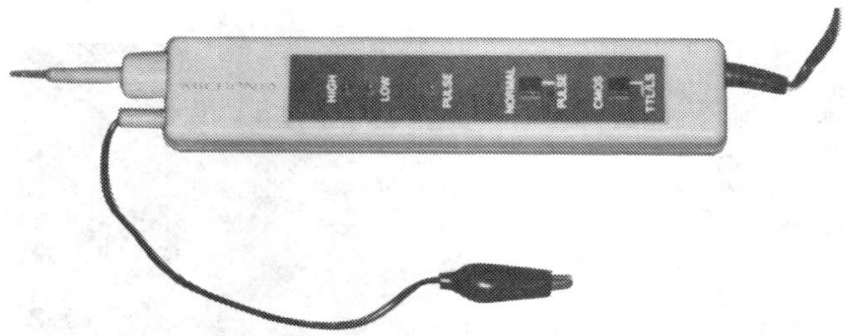

**Figure 6-16.** Logic probe.

operate between 3 and 18 V. Therefore, to use it, set the device-type switch to CMOS or TTL/LS, according to the family of logic devices being tested. Next, set the mode switch to either Normal or Pulse. In Pulse mode, the logic probe will detect the logic level of the pulse train. In Normal mode, the unit will indicate the DC level.

## Using the Logic Probe

Now that you are familiar with the switches, initial setup, and basic characteristics of the probe, you are ready to use it on a circuit. The probe shown is typical. It uses multicolor LEDs and a tone output. It is designed to operate from any power supply ranging from 4.5 to 15 V. After connecting the power leads to a proper supply, touch the metal tip of your probe to the circuit point to be tested. If the low (typically green) LED lights and the unit makes a low tone sound, it indicates a low logic level (for TTL lower than the threshold voltage of about 0.8 V). If the high (typically red) LED lights and there is a high tone sound, it indicates a high logic level (above the threshold level of about 2.2 V). If the circuit to be tested is CMOS, then a low is about 30% of $V_{CC}$ and a high is about 70% of $V_{CC}$.

If no LED is lit and no sound is emitted, the level is either between the logic level threshold voltages or the circuit is "open." If the Pulse (yellow) LED is blinking, it indicates a pulse train at the test point. The duty cycle of the pulse may be qualitatively analyzed by comparing the relative brightness of the high (red) LED and the low (green) LED. If the high LED is brighter, then the pulse has a longer high duration. If the low LED is brighter, the low duration of the pulse is longer.

## Digital Troubleshooting with a Logic Probe

In a combinational logic circuit, the trouble may be located by static gate analysis, that is, by probing the inputs and outputs of the gates under constant DC levels and comparing the results with the truth table for that particular gate.

A gate may appear functional when observed statically but may fail under a dynamic condition (pulsing input signal), which is the actual circuit operating condition. If the circuit is driven by a clock signal or multivibrator (MV), first check the output of the clock or MV circuit; then trace the signal through the circuit.

## Digital Logic Pulser

If the circuit is not driven by a clock or MV series of pulses, a logic pulser may prove useful. Also, if the pulse frequency is faster than you would like in order to step the circuit through a sequence of states so you can check each state with the logic probe, the logic pulse is again the tool for the job. With the touch of a button, a single pulse or a pulse train is delivered to the circuit under test.

The logic pulser in Figure 6-17 is a companion tool to the logic probe in Figure 6-16. It also costs less than $20. By using this pair of tools, you can pulse and monitor different points in the circuitry. This ability to test gates and sequence counters at various points can make it easy to locate a malfunctioning component or group of components.

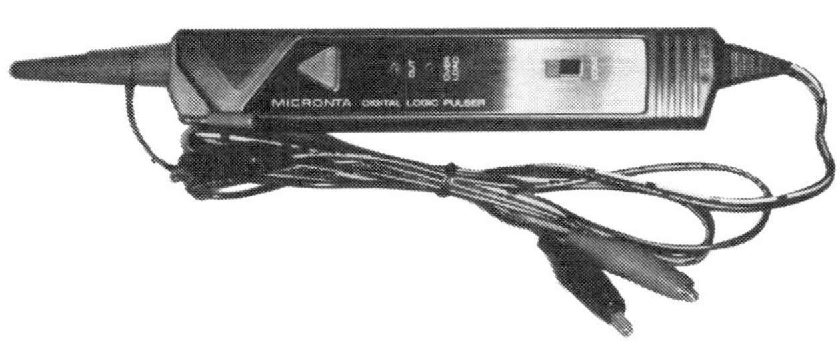


**Figure 6-17.** Logic pulser.

Here are some precautions to observe when using the logic pulser:

1. Test for overvoltage by connecting the power leads to the source. If the source is over 15 V, then the Overload LED will light and a different power source must be used.
2. Test for overcurrent after you have checked the source voltage and determined it to be safe—no more than 15 V. Do this test for overcurrent by touching the tip pin of the logic pulser to the input of the circuit being tested and then pressing the Push button. If the Overload LED does not light, then connect the ground lead to a ground point near the circuit point being tested. If the LED lights during the current overload test, this means that the output current from the tip pin will exceed 200 mA, and you will be unable to insert a pulse at this point in the circuit.

*Note:* When the ground lead is connected, the logic pulser's overcurrent protection circuit is disabled. Do not change circuit input test points without first performing the overcurrent test.

To inject a single pulse at the input of a gate or flip-flop, set the Pulse/Cont switch to the Pulse position, touch the tip pin to the point of the circuit that has been verified to not cause an overload, and press the Push button. A 5-$\mu$s positive pulse will be generated at the pulser tip, which has sufficient output drive to take any normal circuit high momentarily. Because of the low duty cycle, the total energy is limited, which reduces or excludes the possibility of circuit damage. The Out indicator will light when the pulse is generated.

To inject a pulse train into the circuit, set the Pulse/Cont switch to Cont and proceed as before. When the Push button is pressed, a continuous train of 5-$\mu$s pulses at a frequency of 5 Hz will be present at the pulser's probe tip. The Out will blink as long as the pulses are being generated.

## Stimulus-Response Training

The technique of stimulus-response testing is not new. In both audio and RF analog circuits, signal generators have been used to inject a signal and then trace that signal through the circuit with an oscilloscope or an instrument called simply a signal tracer.

In digital troubleshooting, the idea of stimulus-response has not been very common. The logic state could be easily measured, but forcing a state change, especially on a line held low, was more difficult. It usually meant disconnecting the device's input from the circuit to drive the gate's input. Disconnecting the input meant using some method as described in Chapter 2, such as unsoldering and lifting an IC pin from the PCB or cutting a run in the PCB. These destructive techniques have to do with the internal structure of TTL gates. To drive a TTL output (a saturated transistor to ground) requires a great deal of current drive. A continuous high would produce a damaging current that would destroy the TTL gate's output transistor. The logic pulser, by delivering the low-duty-cycle (short duration) pulse of both high current and low energy, makes it unnecessary in most cases to disconnect a driving gate's output. This makes the technique of stimulus-response testing of digital circuits as convenient as it ever was for analog circuitry and greatly reduces the destructive practice of cutting PC runs and unsoldering IC pins for testing.

## The Current Tracer, Clip, and Comparator

Hewlett-Packard has been a leader in developing small testors of digital logic circuits, and it has a family of instruments it calls the "I.C. Troubleshooters." They include the logic probe and logic pulser discussed and a current tracer, a logic clip, and a comparator. The logic probe, logic pulser, current tracer, and logic clip are shown in Figure 6-18.

The current tracer can be used to precisely locate low-impedance faults in digital circuits by sensing the magnetic field generated by fast rise-time current pulses. The logic probe is voltage sensitive and therefore displays voltage information only. The probe indicates if the node was high, low, open, or pulsing. However, if the node was stuck in one state, it may be unable to cross the threshold levels. For example, on a multiple-input circuit as shown in Figure 6-19, the tracer can verify that the driving gate is functioning, and then locate the problem input by tracing the current to the source or sink causing the node to be stuck.

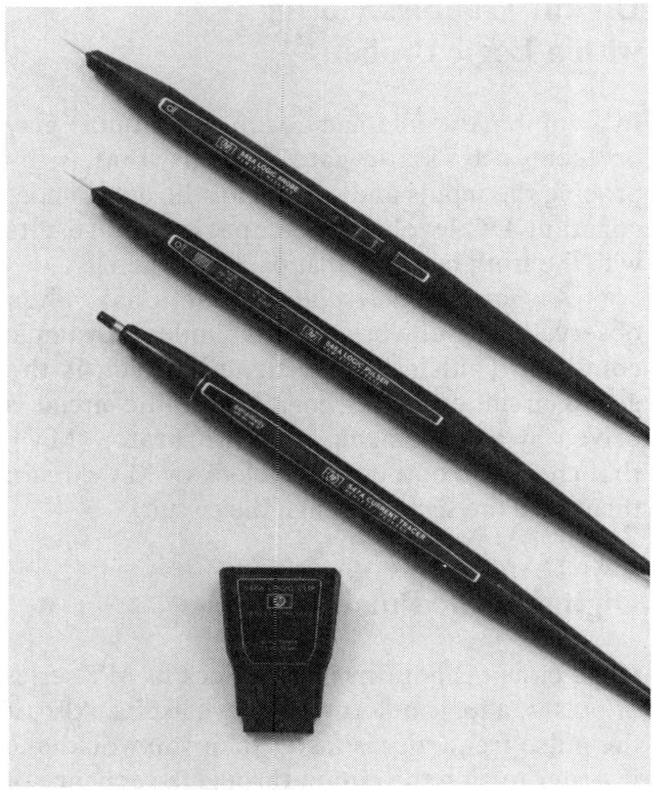

**Figure 6-18.** Logic probe, logic pulser, current tracer, and logic clip. *(Courtesy Hewlett-Packard)*

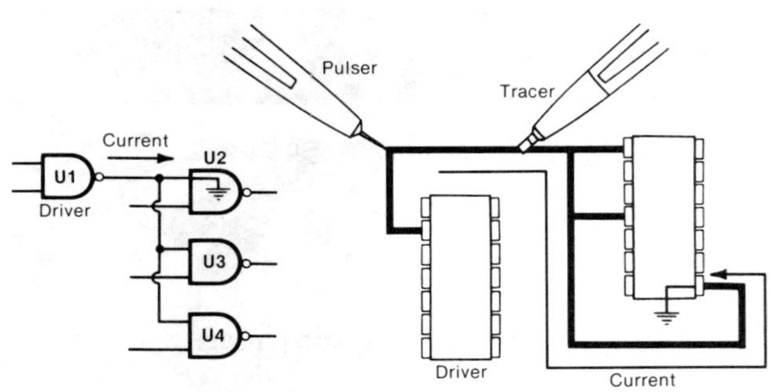


**Figure 6-19.** Multiple-input circuit gate with a shorted input to ground.

Gate U2 in the figure has a shorted input to ground, causing the node to be stuck low and sinking virtually all the current from the driving gate U1 and the other gate inputs. Without the current tracer, each gate input would have to be removed from the node to find the one that was shorted, resulting in possibly inducing more problems in the board.

To use the tracer, simply align the dot on its tip at a reference point, usually the output of the node driver. Set the sensitivity control to indicate the presence of current activity. Then you can trace the circuit to see where the current is. As you follow the PCB traces, probing from point to point, the lamp will change intensity. When you find the faulty input or shorted trace, the lamp will probably indicate the same brightness that it had at the starting point or reference node. Whether the problem is a solder bridge that may be undetectable to the naked eye or a faulty gate with an open or shorted input, the absence or presence of current will lead you directly to the fault.

The tracer has enough sensitivity to follow traces in multilayer boards, through the insulation of a shorted cable, and for tracing faults on computer backplanes and motherboards. So, the current tracer eliminates "shotgun" approaches to digital troubleshooting by showing exactly where digital current pulses are in the circuit, which solves the most vexing troubleshooting problems.

Also shown in Figure 6-18 is the logic clip. This is a multipin state indicator that clips directly onto ICs. The logic state of each pin is displayed by an individual LED, enabling you to check the device's truth table.

## The Logic Clip

Using the logic clip clarifies the relationship of inputs to outputs on a gate, flip-flop, or shift register. For example, just step a counter or a shift register through a complete cycle to verify outputs, resets, clears, and other signals. A pulser like the 546A is a valuable stimulus tool for this type of testing. The clip then quickly shows if a device is following its truth table or if a logic fault exists.

The logic clip in Figure 6-18 is simple to use because there are no controls to set, adjustments to make, or external signal connections to hook up. The clip uses its own internal self-seeking logic circuitry, so it does not matter if it is placed on a device upside down. It immediately locates the supply and ground, and then starts indicating the state of all pins.

## The Logic Comparator

The logic comparator utilizes comparison techniques to identify faulty nodes in a digital circuit. It dynamically tests an IC in circuit by comparing output responses to those of a known good IC, which is plugged into the comparator. Once bad nodes are located, the other members of the IC troubleshooters family are used to locate the specific defective component causing the fault.

Typical setup of the logic comparator would involve locating a source of $V_{CC}$ and a ground on the circuit board for connection to the logic comparator. A logic comparator draws its own small power requirements from the circuit under test, mainly to light up the LED pin indicators. After the logic comparator is connected to the board, the IC clip of the comparator is hooked up to the suspected IC. A known good reference IC is then removed from an antistat case and placed in the logic comparator by means of a zero-insertion force socket. The comparison test is initiated by a "test" push button.

If the IC under test is good, the tester will indicate it. If the logic states do not compare, the logic comparator will help you see why. LEDs adjacent to the known good IC will light at those pins where the logic state of the IC under test is different from the reference IC.

Although the logic comparator is a major timesaver in many applications, it is not 100% effective. When used on high-speed triggered devices, the small propagation delay time of the tester and its cable prevents identical timing in both ICs. Also shift registers and counters must be reset and synchronized with the reference IC, or erroneous test results may occur.

Logic comparators provide a versatile, handheld means to verify the operation of questionable ICs in circuit and under dynamic (powered-up) conditions. However, they are not intended to totally replace standard test equipment, such as the oscilloscope, DMM, logic probe, and pulser. Each has its special applications and benefits.

# Data Communication Testers

When troubleshooting digital systems you may become involved in testing the data communications network, which may seem like an awesome task. However, most tests can be made with small portable units. We will discuss briefly the application of the most common units.

## Breakout Boxes

The *breakout box* (BOB) is considered the standard tool for configuring nonstandard devices to function in a standard environment. It also allows two data terminal equipment (DTE) devices or two data communication equipment (DCE) devices to communicate. BOBs allow the enabling/disabling of leads by jumping across the interface signals, and crossing over and busing of interface signals. One common use of the BOB is to cross pins 2 and 3 of the RS-232 interface. This allows two DTE devices to communicate with each other. A BOB allows you to determine the status of all data, control, and clock signals on the data communication link, at the interface, or at the port.

## Bit Error Rate and Block Error Rate Testers

Bit error rate (BERT) and block error rate (BLERT) testers are versatile devices that combine most functions of all other types of testers in one instrument. They can perform functions such as looping information back, testing the modem circuitry, and sending messages to a terminal. You can select a test signal (message), sync or async, select band rate, loop it back, and then measure the error rate. The BERT involves generating this known data sequence into the transmitting modem, observing the received sequence at the distant modem, and evaluating that sequence of errors.

This type of device can be used to test line, modems, protocols, and anything with an RS-232 interface. If done properly, the system parameters are measured simultaneously such that a total profile of the system performance at a given time can be obtained.

## Status Activity Monitors

Status activity monitors (SAMs) devices allow you to monitor the various leads with LEDs and show the status of signal activity. They also allow you to change lead structures. Many SAMs also function as BOBs, providing both monitoring and modifying capabilities.

## Data Transmission Test Sets

A data transmission test set is a very high performance service and troubleshooting instrument that can locate problems that may elude the BERT or the SAM. It usually contains a BERT, but it also contains built-in diagnostic routines with a keypad entry on the front panel and a programmable mode that allows for automatic transmission, reception, and examination of data. Provisions for storing the data are also included in this test set. Some of its functions include monitoring, analyzing, testing, and troubleshooting a wide variety of data communication interfaces. It can operate as a serial data transmission monitor and as a modem simulator for off-line DTE testing.

Many service technicians are hesitant to use these newer, more sophisticated test instruments, even though they will raise their testing and troubleshooting performance levels.

# 7 The Logic Analyzer

Because of its power as a measurement and troubleshooting tool, the logic analyzer has become the instrument for testing the complex digital systems in today's technology. Both timing analysis and state analysis are presented with specific applications to hardware and software troubleshooting. Many logic analyzers have unique glitch analysis capabilities, which are discussed along with examples of representative manufacturers' instruments. This chapter also compares logic analyzers to oscilloscopes and other microprocessor troubleshooting equipment.

## Overview of the Logic Analyzer

Although a multichannel oscilloscope can be used to observe general signal activity on the various microcomputer bus lines and gives much better visibility into an analog circuit than a voltmeter, the logic analyzer provides much better visibility into a digital circuit than does an oscilloscope. The oscilloscope is still an essential tool for observing and measuring voltage parameters at any given point in a circuit in a time domain. In fact, it has been said jokingly that, "There is no such thing as a digital signal; digital is just a fast analog."

The evolution of circuitry from the single-point measurement of analog to the multipoint measurement of digital demanded a corresponding evolution in test equipment. To meet this demand, instrument manufacturers have developed a bewildering array of new instruments to expedite the troubleshooting and repair of digital/microprocessor equipment. Among these new instruments, the one that stands out as most versatile and effective in microprocessor-based systems is the logic analyzer. It can increase the proficiency and productivity of the troubleshooter and minimize time and material cost associated with digital repair. The logic analyzer enhances the vision of the digital troubleshooter, which allows the defect to be located more rapidly.

The logic analyzer provides substantial saving in material costs by helping isolate a problem to an individual component. Quickly and reliably isolating problems to the component level will make board swapping less frequent. Getting PCBs repaired and back in service means there is less valuable inventory tied up in replacement boards and less wasted time waiting for a board exchange. Increased productivity can be immediately substantial.

The logic analyzer is extremely useful in microprocessor testing. It is the only device that can be used to view all of the timing of a microprocessor functioning in a system. It has multiple-input probes that connect its many channels to various points in the microprocessor circuit under test. Logic levels of the circuit under test are interpreted as highs or lows and continuously placed into a high-speed RAM memory. The special high-speed memory used in logic analyzers is operated like a giant shift register, taking data in at one end and dumping it out the other end in a bit bucket to be thrown away.

When the logic analyzer is triggered, it stops acquiring data. This data in the memories can be displayed as timing diagrams, shown as binary 0s and 1s in a state table, recorded on a diskette for future reference, or treated in various other ways.

## The Logic Analyzer Compared to the Oscilloscope

The oscilloscope, for the most part, is a real-time measuring instrument. Exceptions are the digital storage oscilloscope (DSO) and the sampling osilloscope. The greatest number of channels that can be displayed on an oscilloscope is eight; more commonly, it is four or even two. A microprocessor-based system may easily require between 24 and 72 channels to be observed simultaneously. This is where the logic analyzer excels. Not only can certain logic analyzers display many channels, but they can also acquire data on these channels simultaneously. Since an oscilloscope is real-time, it must time-share between channels in an alternate or chop mode to acquire any data for its eight or less traces.

All logic analyzers capable of displaying a timing diagram must display a pseudowaveform, because they never acquire the actual voltage level for each point in time. Instead, most analyzers, as indicated, simply determine whether the signal is above or below a given threshold voltage.

Since the oscilloscope displays in real time, an advantage it has over the logic analyzer is the ability to display the actual signal amplitude as a function of time. Some oscilloscopes today can acquire, store, and display signal widths of less than a nanosecond. Modern logic analyzers are not that fast—they cannot acquire nor resolve time intervals less than or even equal to a nanosecond.

A powerful feature of the logic analyzer that is found in a few DSOs is the ability to trigger on a preassigned digital word or pattern of 1s and 0s. This feature is called *word recognition*. Coupled with word recognition is the ability to position the acquired data so that events before (pretrigger), after (posttrigger), or both before and after (center trigger) may be viewed. This allows occurrences that led up to the trigger fault to be seen.

One of the most important advantages that the logic analyzer has over the oscilloscope is the display format available. Besides the timing diagram display, the logic analyzer has the capability of displaying the data in binary, octal, hexadecimal, decimal, mnemonics, ASCII, GPIB, map, or signature analysis (see Chapter 9).

## Types of Logic Analyzers

Several types of logic analyzers have a wide range of capabilities. Some instructions combine several functions, making them multifunction test tools. Some of the functions and/or types found are

General-purpose (interactive)
Serial

State analysis
Microprocessor
Analog analysis
Performance analysis

The general-purpose analyzer can be connected to various points in the circuit and handle more than one logic level voltage such as TTL, HTL, or CMOS. Most general-purpose logic analyzers can operate with an internal or external clock. They will always include a timing diagram and/or state table, and may include other displays. The data can be acquired in parallel channels with timing between channels compared. One of the channels can be viewed to look at serial timing information and may even contain one or more digitized analog channels, making it also a DSO.

The serial analyzer is necessary to do a serial-to-parallel conversion on serial data for a meaningful display and flexible triggering. Displays will vary, but hexadecimal and ASCII are common.

Some so-called logic analyzers are little more than multiplexers, made as add-on adapters for an oscilloscope or a personal computer. They provide only time waveform diagrams, usually with wide magnification range. Sometimes, they have user-definable labels.

State analysis listing is available only on instruments called state analyzers. They generally provide displays and windowing of address, data, status, and control line activity. Display modes are selected from a list such as binary, octal, decimal, hexadecimal, ASCII, and user-defined mnemonics and/or microprocessor specific mnemonics. A logic state analyzer has no internal clock, so an external clock must be used for acquiring data. This is not a disadvantage as long as you are interested only in data that occurs at the system clock edge.

The microprocessor analyzer is an instrument ideally suited for use on a microprocessor-based system. It has 24 or more input channels to acquire data on the address bus, data bus, and control lines. A clip to put on the microprocessor socket may be the only source of connection, though there may be additional general-purpose channels available. Displays on the microprocessor analyzer will vary but will usually include disassembled machine code mnemonics. This requires that the analyzer have a microprocessor-specific personality card or module for each different microprocessor for which it is used.

Analog waveform analysis is provided on some instruments simultaneously with a display of up to two channels that have user-definable labels, a wide magnification range, and direct readout of time and voltage cursors.

System performance analysis capabilities provide time interval and event histograms that allow viewing system hardware/software activity or specific modules for performance evaluation. Out-of-specification conditions or bottlenecks in program flow or in hardware execution stand out. The display also shows measurement data, including minimum, maximum, average value, and total measurement time.

Each of these functions and types of logic analyzers will be discussed in more detail.

## Digital Domain

We live in a time-base continuum. Therefore, our earliest measurement experiences were those in a time domain. The general-purpose oscilloscope has been the mainstay for this type of measurement. When electronics applications expanded to include frequency as an independent variable, a measurement concept evolved in the frequency domain—the spectrum analyzer entered as a measurement tool. Recently, the concept of the data domain has appeared in order to describe measurements made in the language of computers. However, the language with which computers process information is insufficient to describe all that transpires in a digital system. The complexities of measurements within a digital system where we find both data language and timing variables require analysis and measurement in a digital domain.

A microprocessor system, such as those discussed in Chapters 3 and 4, contains all of the characteristics of the digital domain. First, the digital domain, whether formatted in parallel or serial, is multichannel. Our microprocessor system might have in parallel 16 or more address lines, 8 or 16 data lines, several control lines, clocks, and other control functions, each carrying some element of information. Simultaneously these buses are interconnecting programs in ROM, operational memory in RAM, and various I/O interfaces. The information itself, however, is only bistate. In this digital world, we are only marginally concerned with absolute voltage levels. We want to know if the information channel is above or below some predetermined threshold, as seen in Figure 7-1. If above the threshold, our information state or logic level is high or 1. If below, the logic level is low or 0.

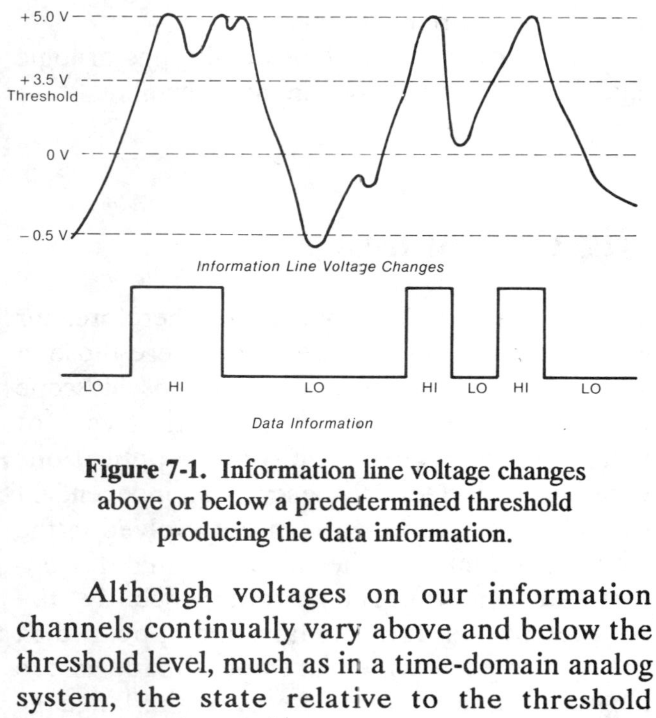


**Figure 7-1.** Information line voltage changes above or below a predetermined threshold producing the data information.

Although voltages on our information channels continually vary above and below the threshold level, much as in a time-domain analog system, the state relative to the threshold becomes information or data only at periodic sampling intervals. Since the sampling of logic states is normally synchronized with one edge of the timing or clock pulse, the speed with which data are processed is referred to as its *clock rate.* However, whether this pulse is uniformly clocked or is random, the status of the information channels becomes data only when this pulse authorizes the machine to acknowledge it. Figure 7-2 shows acquiring data at a clocked interval.

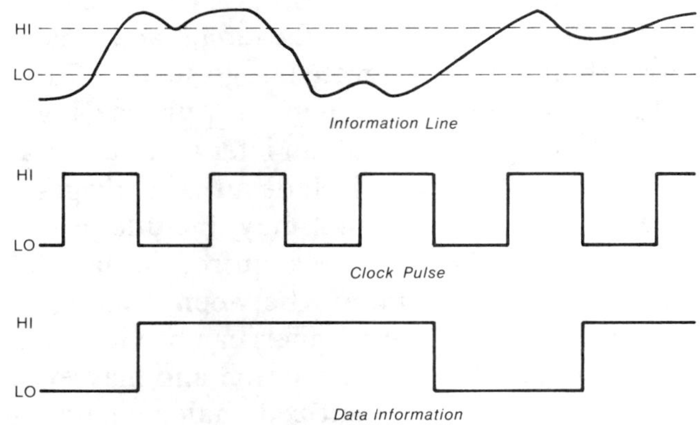


**Figure 7-2.** Acquired information becomes data only at periodic sampling intervals synchronized with the clock edge.

Another characteristic of the digital domain is that it is essentially single-shot. A given condition, whether error condition or valid condition, occurs only once in a given routine, or occurs so seldom or aperiodically that it appears single-shot when viewed from a time domain.

Also inherent in this multichannel data world is the possibility that a data anomaly may be the only symptom of a prior electronics malfunction or may be the result of some unexpected prior data sequence. A general-purpose oscilloscope uses its trigger to initiate its sweep, making it a powerful tool for looking at subsequent information. However, most digital fault indications are symptoms of previously occurring problems. An oscilloscope can display only what occurs after the symptom. The major challenges to the oscilloscope in the digital world are its inability to display in negative time and its inability to handle adequately many information channels simultaneously.

## Data Acquisition

The logic analyzer has a data storage system that continually retains data, so when an error trigger occurs the preceding data have already been captured. This is done by holding data in a sequential access memory, such as a shift register.

## Comparator

A block diagram of a general-purpose logic analyzer is shown in Figure 7-3. The individual probes of the lead set to each parallel channel are connected to comparators. A single-threshold comparator will output a high if the input is above the threshold selected by the user. It will output a low if the input is below that threshold. This high or low is what is stored in the acquisition memory, not the actual voltages of the input signals.

## Probe Pod

The input channel comparators are usually located inside the logic analyzer or in the probe pods. Probe pods are more convenient because there are fewer connections to the logic analyzer from the circuit under test than with individual test probes. A probe pod is shown in Figure 7-4A, and several probe pods to connect to a microprocessor circuit are shown in Figure 7-4B. Each connected probe has a POD ID button. If the button is pressed, a readout of the probe's pod ID is displayed on the logic analyzer's screen. This readout can be used to verify that the probes are connected properly.

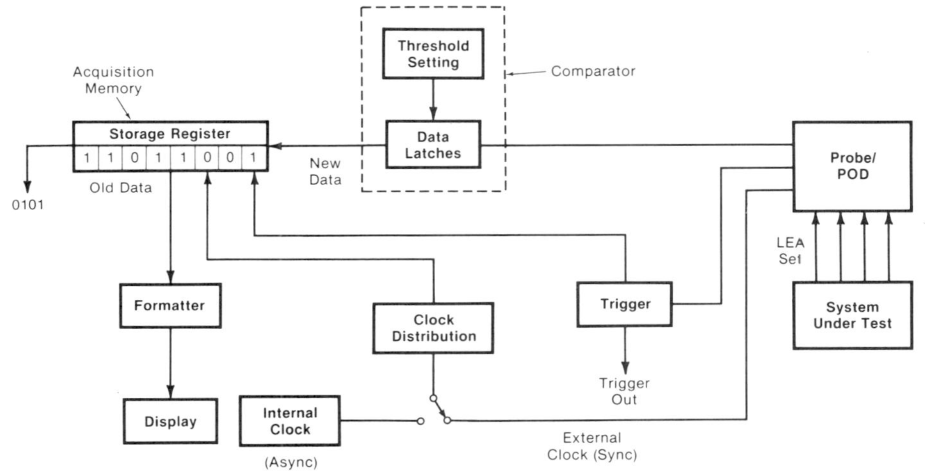


**Figure 7-3.** Block diagram of a general-purpose logic analyzer.

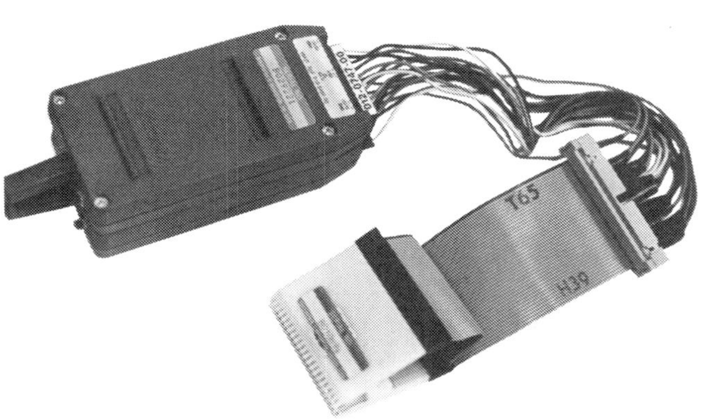

(A) Logic analyzer probe pod. Note the POD ID button.

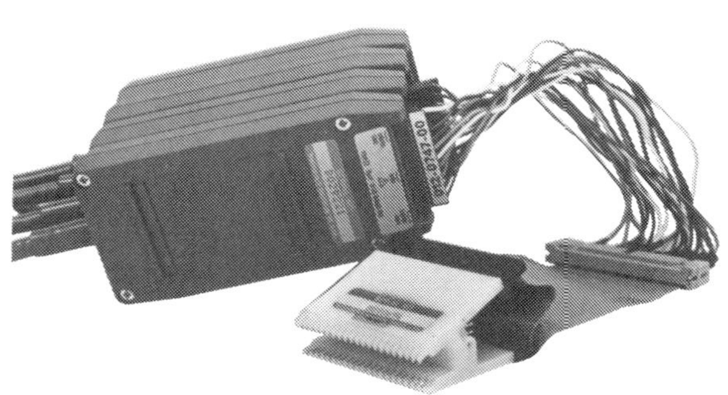

(B) Several probe pods to connect to a microprocessor circuit.

**Figure 7-4. Probe pods.**

## Synchronous/Asynchronous

Each bit of data is introduced into the storage register with the transition of a clock pulse. As we saw in Figure 7-2, the continually changing voltage levels on a multitude of lines become data only at the instant that the appropriate lines are acknowledged.

As one approach to logic analysis, we might want to synchronize our perception of data with that of the system under test. In this case, we choose to synchronize to a clock that is external to the logic analyzer and synchronous to the system being monitored. This is referred to as *synchronous data acquisition*. This is the most significant feature of that class of logic analyzer called logic state analyzers and is the essence of the data-domain concept. With synchronized data clocking, we get a macrolevel view of the digital system as it is supposed to work. We see, for instance, whether the microprocessor system recognizes instructions in the sequences that the programmer intended. Logic state analysis can quickly reveal an invalid logic state. It may not, however, reveal the cause of the invalid state, particularly if it is caused by an electrical anomaly such as a slow switching device. Here we need a microlevel look at the timing relationships between the various information channels as they change logic states. This becomes feasible if we divorce our perception of data states from the system's clock and substitute a separate clock that is internal to the logic analyzer to acquire this data. This auxiliary clock can easily be generated within the logic analyzer. Data processed by the logic analyzer are therefore asynchronous with the microprocessor system.

As a practical matter, an asynchronous clock may be five times as fast as the system clock, which gives good resolution of the timing sequence in which data states change. It is this resolution that differentiates another class of analyzers—the logic timing analyzer. Our digital domain measurement system, therefore, accepts data into the acquisition memory by sampling the information channels either synchronously with the microprocessor system's clock or asynchronously at a faster rate. Or we might want both clocking capabilities.

## Clock Qualifier

Often a digital bus in the microprocessor system will not only have data of interest but will also have irrelevant data as well. For example, we might be interested in data on a data bus during memory read but not during memory write, or vice versa. There are always one or more control lines in the system under test that will tell the system when there is a memory read. A logic analyzer with a clock qualifier can gate the external synchronous clock through the location with the selected control line. This saves memory space in the logic analyzer.

## Digital Latch

A *glitch* is a pulse narrower than the minimum pulse duration and/or more than one transition between clock edges. A digital latch may act as a glitch catcher; it simply latches very narrow pulses that a logic analyzer would otherwise miss. The status glitch(es) or no glitch is stored in a glitch memory and can then be indicated on the display.

## Memory Configuration

We have seen that the digital world is multichanneled. This is handled by simply stacking the input gates, comparators, and memory registers in parallel. Our logic analyzer may employ a fixed number of channels. To provide flexibility for different data needs, the number of inputs may vary. There could be 4, 8, 16, 32, or more inputs. Input flexibility is much more significant than the maximum input capability in terms of probes, gates, or comparators. You can always choose to use less than the maximum capability, if you do not need that many channels. However, you may not want to waste any memory capacity. If a given analyzer system has a total storage register of 4096 data bits and it can handle 16 parallel inputs, then 256 data bits can be stored in each channel. Now 256 words is a good long data stream to analyze if you are looking at 16-bit words. However, if you are looking at only 4 channels, each 256 bits long, three-fourths of the logic analyzer's memory capacity would be wasted. That capacity might be needed if you were examining long data trains with a very fast asynchronous clock for close timing analysis. A far more versatile digital domain instrument would result if the storage capacity could be formatted; that is, if the total capacity could be divided among the selected number of input channels. For example, our 4-kb analyzer could be formatted to store and display 256 data words of 16 bits each, 512 words of 8 bits each, or 1,024 words of 4 bits each. So, with multichannel inputs and formattable memory, our analyzer would be much more adaptable.

## Triggering

In an oscilloscope, the trigger event starts the sweep and, therefore, the display of information. In a logic analyzer, the trigger event stops the data acquisition process. When the trigger event occurs, telling us that our storage register is filled with data of interest, data input ceases, thereby freezing the contents within the register. Data in the storage register can then be transferred to the analyzer display for use by the operator.

### Pretriggering and Posttriggering

Although the capture of data fulfills our need for data storage and for storage of data prior to a trigger, it gives us a view of data that is too limited for practical use. We may want to view some of the data leading up to and following the trigger. As shown in Figure 7-5, the data before the trigger are called *pretrigger* and the data after are called *posttrigger*. At times, we might want only a small amount of data either before or after the trigger; or we might want a balanced amount of data on either side of the trigger. In practice, when the trigger event is sensed, the storage register continues to accept data for a few more clock pulses. A posttrigger counter continues to pass clock pulses up to a variable count equal to a percentage of the storage register's capacity. The register then stops receiving new data. The resulting display is divided between data occurring pretrigger followed by posttrigger data. The balance of data on either side of the trigger is controlled by the setting of the variable counter.

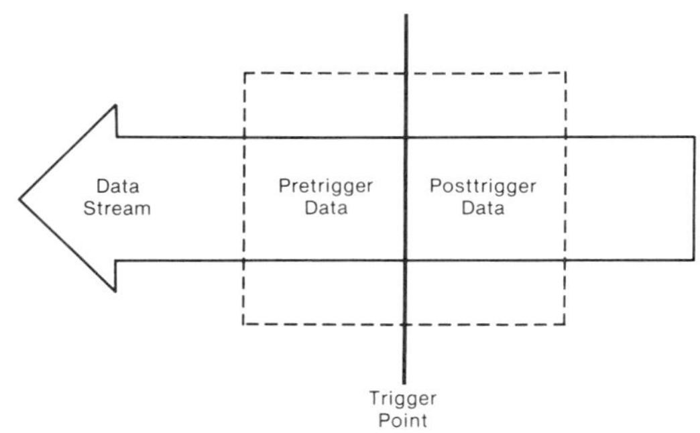


**Figure 7-5.** Effect of the trigger event on the data stream; pretrigger and posttrigger data.

### Digital Delay

Another option for digital-domain triggering is digital delay. Occasionally you may want to capture a block of data that occurs many clock cycles downstream from the trigger. The output of the digital delay can be set to count a given number of clock cycles after the trigger word and then stop the data acquisition process.

Another use for digital delay is to generate a trigger, not on the first occurrence of the trigger event, but on the 2nd, 3rd, or 100th. The digital delay would be set to count the appropriate number of trigger events, then output a trigger to stop the data acquisition process.

### Word Recognition

A trigger might be a rising or falling edge on a single unit, such as is often used to trigger an oscilloscope. For a logic analyzer, a more useful trigger is a combination of events. An example is an address bus in a microprocessor system that will have various combinations of 0s and 1s for different addresses. A *word recognizer* is a circuit that produces a trigger when the combination of 0s and 1s at its input matches a specific configuration. A word recognizer is equivalent to a large AND gate. If the user wants to match low (0) inputs rather than high (1) it will have inverted inputs on each channel. If the application requires triggering on events that are not synchronized to a system clock, as seen in Figure 7-6, the word recognition is asynchronous.

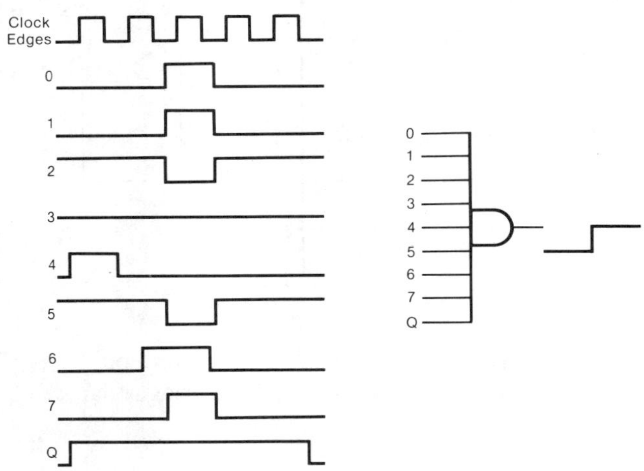


**Figure 7-6.** Asynchronous word recognizer with an additional word qualifier input.

An address bus or other word recognizer input might coincidentally have the right combination of 0s and 1s between the clock pulse edges or at a time other than that of interest. Synchronous word recognition, word qualifier, and word recognizer filters are techniques that are used to make the trigger event more unique and more selective.

Synchronous word recognition causes a trigger condition only if the trigger event (proper combination of inputs) occurs only on a system clock edge. A small addition to the AND gate makes possible this ability to synchronously word recognize.

It is common to word recognize and acquire data for display from the same lines on the system under test. However, there are applications where one or two additional lines are needed for word recognition that need not be acquired and displayed. These lines are referred to as ***word qualifier inputs***. Like the external trigger input of an oscilloscope, these qualifier signals are not acquired for display.

In an asynchronous microprocessor system under test, narrow glitches might be present but ignored by that system. Glitches between clock edges are ignored in synchronous word recognition. These glitches could, by coincidence, produce the right pattern of 0s and 1s that could be recognized in asynchronous word recognition. By extending the rise and fall times of each line prior to the AND gate, these glitches can be effectively filtered out of the asynchronous word recognition process. Because the word recognition filter is needed and used only in asychronous word recognition, it is often referred to as an *asynchronous filter*.

### Range Recognition

Range recognition can be useful in finding problems in a program. A program stored in a particular address range may jump out of that range for no apparent reason. A range recognition feature can generate a trigger if any address (trigger word) outside of a particularly specified address occurs. The opposite situation, that of entering a particular address range, can be set up to cause a trigger within that range.

## Types of Displays

We have seen that we need to acquire many channels of data, continually store this data, and then trigger our system to stop storing at the appropriate time. We now look at how we might want to format our information for display and use. The availability of a variety of displays ensures ease of use, less time for data interpretation, and fewer errors.

### Timing Diagram

To look for timing sequences and propagation delay times, a timing diagram display is appropriate. Figure 7-7 shows a typical example of such a timing diagram. Notice that the waveforms are very rectangular and clean looking. The waveform in memory for each point in time that was sampled is simply put on the display as a high, or low, horizontal line. The formatter may draw a vertical line to connect any transistion from a high to a low, or low to a high. Because the display is a pseudovoltage waveform, information such as ringing and rise times never gets into the logic analyzer memory and therefore does not appear in the waveform. Rather than time, the horizontal axis represents a sequence of clock increments with state levels represented in clock period increments.

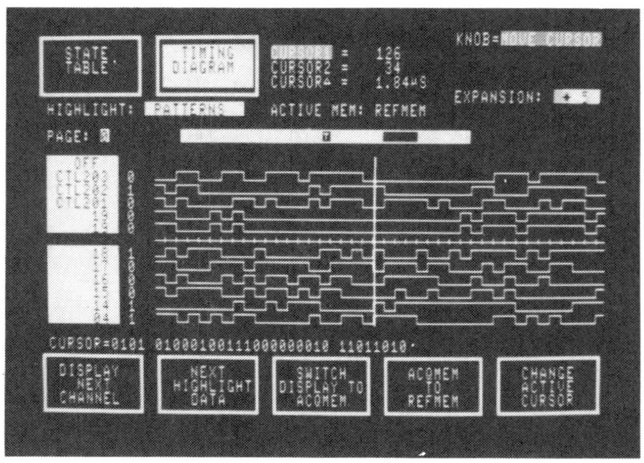


**Figure 7-7.** Timing diagram display.

To aid in comparing many lines on a timing diagram, we would probably want them spaced in convenient groupings, be able to expand them both vertically and horizontally, and shift the display along both axes for closer examination at selected points. We also would want some indication on the display to quickly identify which data word is the trigger word.

## State Table

Many data state troubleshooting problems are more conveniently displayed in comparison to a programming truth table. The stored data can then be converted to 0s and 1s and arranged in parallel data words on a binary state table, as shown in Figure 7-8. The binary state table shows the binary value of each individual channel for a word acquired in parallel. Channel 0 is the right column. For convenience in converting to binary-coded decimal or hexadecimal, group columns in 4 bits each. If octal notation is more convenient, group the columns into 3 bits each. The logic analyzer can even format the words directly into octal or, as shown in Figure 7-9, hexadecimal mnemonics. A few logic analyzers, like that in Figure 7-10, are capable of these other combinations of displays.

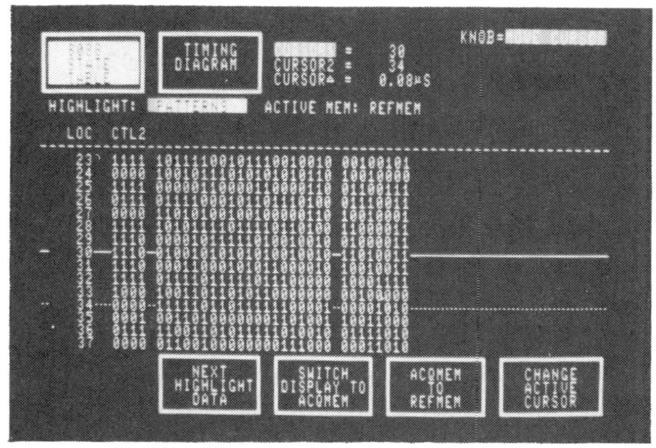


**Figure 7-8.** State table diagram display in binary.

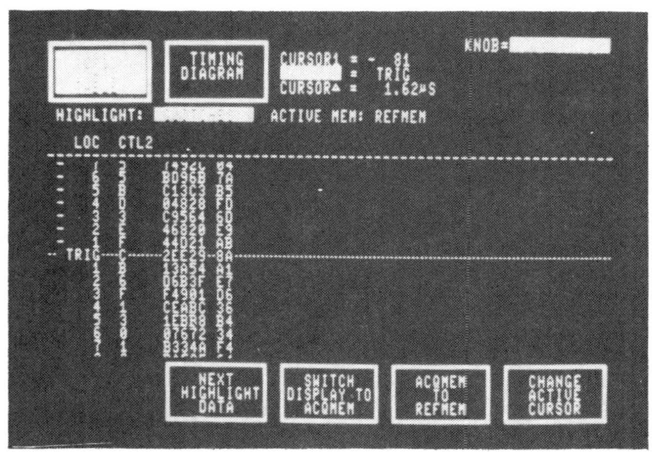


**Figure 7-9.** State table diagram display in hexadecimal.

## Mnemonic Display

A mnemonic display is useful for those who work with assembly level languages. ROM packs are

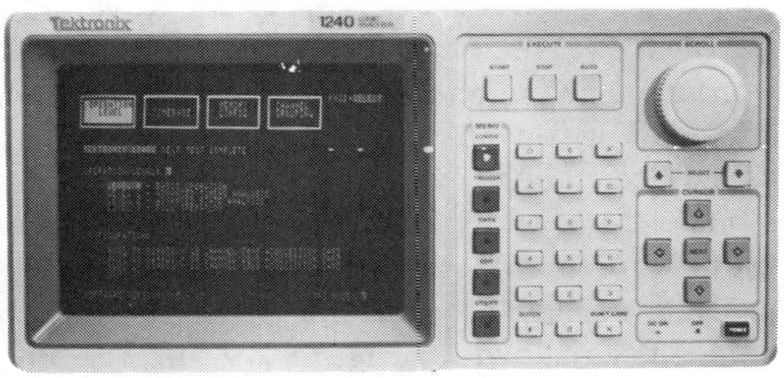


**Figure 7-10.** The 1240 logic analyzer.

available that provide the capability to capture data that are very specific to a problem and assist in analyzing that data by processing it and presenting it in the most useful manner. These ROM packs support mnemonic disassembly of most popular microprocessors. For example, JSR (jump to a subroutine) is less confusing, less prone to error, and much more meaningful than just 8D, the hexadecimal equivalent for a 6800 microprocessor. Figure 7-11 shows an example of a display that contains mnemonics. These features of special display formatting are especially valuable to the troubleshooter who might not be as comfortable with the machine and assembly languages. The logic analyzer's ROM pack and formatter convert to the more common displays to save the troubleshooter time and possible errors.

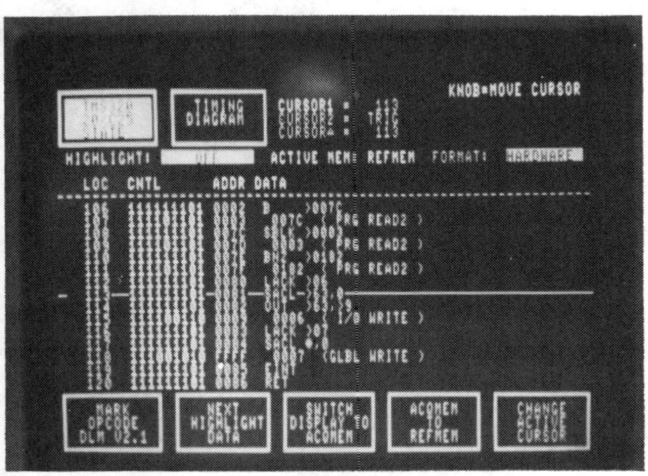


**Figure 7-11.** A logic analyzer display containing disassembled mnemonics for a TMS32020 digital signal processor.

## Map Display

Another method of formatting logic states is a map or state matrix. Coordinates are established along both horizontal and vertical axes, so each point value on the matrix corresponds to a unique binary word. As the analyzer steps from word to word, program flow is shown. The technique is shown in Figure 7-12. An 8-bit word, for example, from the address bus divides into 4-bit halves. Each half then goes to a digital-to-analog converter, then to the vertical and horizontal amplifier, respectively, of an X-Y display devices. This is a useful troubleshooting or analysis tool to isolate trouble spots in memory or to analyze a program sequence for improvement in efficiency.

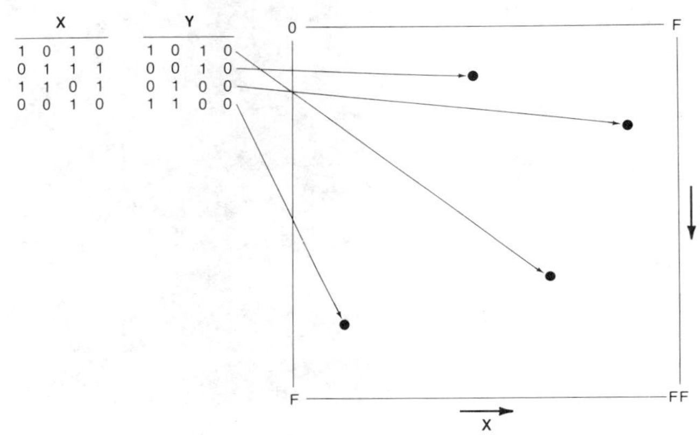


**Figure 7-12.** Map display.

## GPIB

An option on some logic analyzers, including the one shown in Figure 7-10, is full listener/talker on the GPIB or IEEE-488 standard interface capabilities. When GPIB lines, such as attention, end-or-identify, service request, and remote enable, are asserted, they are indicated on the display by ATN, EOI, SRQ, and REM. These acronyms on the GPIB data bus are more meaningful than 0s and 1s. Since timing on the GPIB handshake lines is important to the proper operation of the bus, it is beneficial to acquire and display the handshake lines as a timing diagram.

Since ASCII is a common, but not required, code for the GPIB, the data can also be shown both in decoded ASCII and hexadecimal on some logic analyzers.

## Using The Logic Analyzer to Solve Digital System Problems

To illustrate locating a fault in a digital circuit, we look at a simple counter that produces the data pattern in Figure 7-13. Each data line is half the frequency of the one before, producing a count from 0 to FF in hexadecimal or 255 in decimal. If 00 in hexadecimal is entered into the logic analyzer as a data word trigger, we can trigger on the beginning of the data word pattern. The data will be displayed posttrigger or after the trigger word position. A start key, or something similar, will begin the acquisition of data. A trigger will occur and be indicated when the trigger word is sensed. This causes the acquisition to cease and storage and display to occur.

The logic analyzer displays a portion of the stored data that can be selected by the display position control. This position control is an up/down key on some logic analyzers and a scroll knob on others. A cursor can be placed on the screen and a readout of the cursor position relative to the trigger word, as well as a readout of the value of the word at the cursor point in the number format, can be selected.

The window size is selected for improved resolution of the data displayed. Figure 7-14 shows a small window selection for the display. Use of the position, cursor, and window controls allows examination in detail of all parts of the timing diagram in memory. In Figure 7-14 the cursor is positioned over a single bit on line 7. It is 64 clock cycles after the 00 word in the pattern. If the pattern were correct, that bit would have remained a 0 instead of changing to a 1.

Data in memory can be examined at any time in timing or state mode by simply pressing

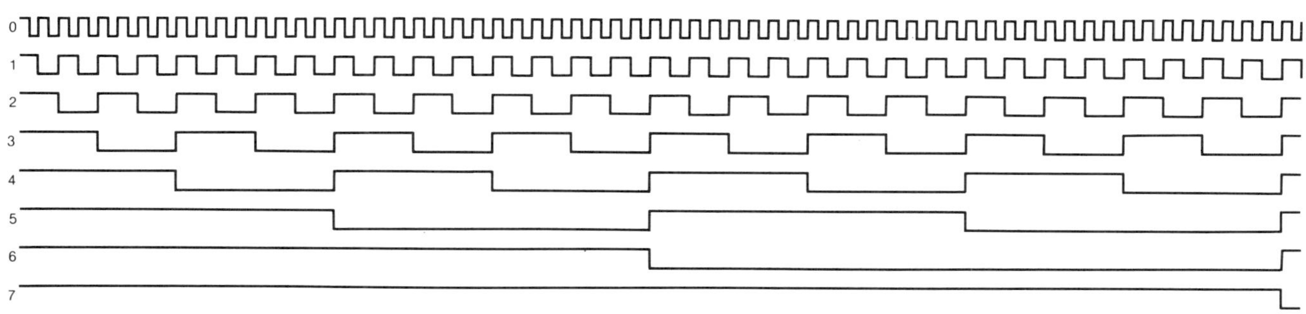

**Figure 7-13.** Counter data pattern.

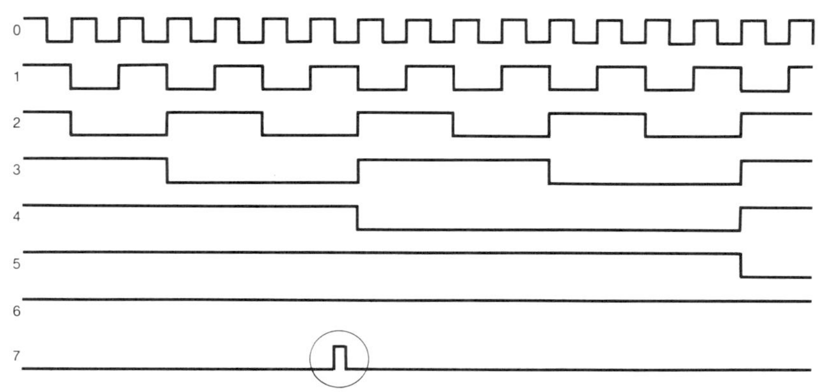

**Figure 7-14.** Counter data pattern expanded.

the mode key desired. In the timing mode, the cursor was positioned at the 64th clock pulse after the trigger word. If the parallel state key is pressed, the state table starts at the cursor position displaying the data in the number formats desired. Figure 7-15 shows the state table with the display presented in hexadecimal, binary, and octal formats. Notice that the circuit is producing a pulse pattern at the cursor position that should logically be a hexadecimal 40 instead of the C0 found there. You might wonder at this point if a 40 occurs in the pattern. The logic analyzer can search for such a pulse pattern by setting up the search mode to look for the occurrence of the searched-for word, in this case 40. The displayed portion of memory then shifts to that area that includes the searched-for word, as well as the positions immediately after, and displays a readout of the position of the word relative to the original trigger word. You can then interrogate the logic analyzer as to whether there are other occurrences of the data word, 40. An analysis of the counter circuit should yield the cause of the malfunction, which can be cleared and then rechecked.

| HEX | 7 | 6 | 5 | 4 | 3 | 2 | 1 | 0 | OCT |
|---|---|---|---|---|---|---|---|---|---|
| CR = C 0 | 1 | 1 | 0 | 0 | 0 | 0 | 0 | 0 | 3 0 0 |
| 4 1 | 0 | 1 | 0 | 0 | 0 | 0 | 0 | 1 | 1 0 1 |
| 4 2 | 0 | 1 | 0 | 0 | 0 | 0 | 1 | 0 | 1 0 2 |
| 4 3 | 0 | 1 | 0 | 0 | 0 | 0 | 1 | 1 | 1 0 3 |
| 4 4 | 0 | 1 | 0 | 0 | 0 | 1 | 0 | 0 | 1 0 4 |
| 4 5 | 0 | 1 | 0 | 0 | 0 | 1 | 0 | 1 | 1 0 5 |
| 4 6 | 0 | 1 | 0 | 0 | 0 | 1 | 1 | 0 | 1 0 6 |
| 4 7 | 0 | 1 | 0 | 0 | 0 | 1 | 1 | 1 | 1 0 7 |
| 4 8 | 0 | 1 | 0 | 0 | 1 | 0 | 0 | 0 | 1 1 0 |
| 4 9 | 0 | 1 | 0 | 0 | 1 | 0 | 0 | 1 | 1 1 1 |
| 4 A | 0 | 1 | 0 | 0 | 1 | 0 | 1 | 0 | 1 1 2 |
| 4 B | 0 | 1 | 0 | 0 | 1 | 0 | 1 | 1 | 1 1 3 |

**Figure 7-15.** State table displaying the data in hexadecimal, binary, and octal.

The compare display mode is used to compare acquired and stored data with data in a reference memory. The reference memory is filled by placing presently stored data within it with a front-panel control key. The logic analyzer can then constantly monitor a stream of data comparing it with a known good, or at least a known, pattern. If any change occurs, the acquisition stops and the different pattern remains in the main acquisition memory. The compare function is then used to compare the two patterns and examine the differences, which are usually highlighted or shown in reverse video. An example is Figure 7-16, which indicates that the LSB is stuck on 0. A quick way to confirm that the channel 0 bits are indeed all 0s is to search for a counterexample. Select the entry format to be binary and the Search mode. Enter X's (don't cares) in all places except the LSB. Since the only data you are interested in is the 0 bit, you are looking for any examples where that bit is a 1. The logic analyzer display immediately tells you that there are no such examples. This in turn indicates that the line is shorted low (to ground). A look at the timing diagram and scrolling through the data would also show the 0-bit line always low. Clear the suspected grounded line by the techniques and methods in Chapters 2, 6, and 10.

| HEX | 7 | 6 | 5 | 4 | 3 | 2 | 1 | 0 | OCT |
|---|---|---|---|---|---|---|---|---|---|
| 4 0 | 0 | 1 | 0 | 0 | 0 | 0 | 0 | 0 | 1 0 0 |
| C 0 | 1 | 1 | 0 | 0 | 0 | 0 | 0 | 0 | 3 0 0 |
| C 2 | 1 | 1 | 0 | 0 | 0 | 0 | 1 | 0 | 3 0 2 |
| C 2 | 1 | 1 | 0 | 0 | 0 | 0 | 1 | 0 | 3 0 2 |
| C 4 | 1 | 1 | 0 | 0 | 0 | 1 | 0 | 0 | 3 0 4 |
| C 4 | 1 | 1 | 0 | 0 | 0 | 1 | 0 | 0 | 3 0 4 |
| C 6 | 1 | 1 | 0 | 0 | 1 | 1 | 1 | 0 | 3 0 6 |
| C 6 | 1 | 1 | 0 | 0 | 1 | 1 | 1 | 0 | 3 0 6 |
| C 8 | 1 | 1 | 0 | 0 | 1 | 0 | 0 | 0 | 3 1 0 |
| C 8 | 1 | 1 | 0 | 0 | 1 | 0 | 0 | 0 | 3 1 0 |
| C A | 1 | 1 | 0 | 0 | 1 | 0 | 1 | 0 | 3 1 2 |
| C A | 1 | 1 | 0 | 0 | 1 | 0 | 1 | 0 | 3 1 2 |

**Figure 7-16.** State table displaying the data in hexadecimal, binary, and octal with the LSB stuck on 0.

Let's look at another example of using the logic analyzer to troubleshoot a different fault. Using the compare function, scroll through the data to see if there is a difference between that in acquisition memory and reference memory. Any differences will be highlighted or shown in reverse video. For our example, data lines 4 and 5 show a difference between the reference pattern and the tested pattern. Bits 4 and 5 seem to always have the same data on both of them. The logic analyzer can tell us if that is true, which will indicate that those two lines are shorted together. So, we look for an example where they are different. In the Search mode again, with binary format, set lines 7, 6, 3, 2, 1, and 0 to X, a 0 in bit 5, and a 1 in bit 4. The logic analyzer says there is no example like that. Try the reverse example—a 1 in bit 5 and a 0 in bit 4. Again, there is a negative result. There are no data where these two lines are different. A glance at the timing diagram in Figure 7-17 will verify that the two lines are shorted. This fault can also be cleared by the methods of Chapters 2, 6, and 10.

## Hardware Performance Analysis

The logic analyzer is of tremendous value in the analysis of digital hardware in the test environment. It can be used for both functional and parametric aspects of digital hardware testing. Functional testing is where the component is checked to verify that it performs its required function, whereas parametric testing verifies that it meets its maximum and minimum specifications.

### Glitch Testing

When digital circuit boards are tested, glitches are often encountered. Glitches are pulses that are narrower than the minimum-pulse-duration criteria or more than one transition between clock edges. Glitches are often symptoms of a circuit problem, usually timing. To fully analyze the problem, the logic analyzer must be able to trigger on the glitch and then isolate when and where it occurred.

### Parametric Testing

A typical application of parametric testing is to verify the threshold sensitivity of a device specified by the vendor. These specified limits need to be verified because the circuit performance can partially degrade when it is operating on the boundary limits. To perform this test, the standard criteria are stored in the reference memory and the test data are stored in the acquisition memory. The two are then compared, and any mismatched data are highlighted. The threshold values of the input test data are then changed and reacquisition is initiated, again comparing the acquired data with that in reference memory. This process is repeated until the test data fail to match the reference data. The thresholds are then compared with the manufacturer's specifications. To examine the threshold values and determine the rate of performance degradation rapidly, the logic analyzer in Figure 7-10 has an auto run test setup feature.

## Software Analysis

The logic analyzer is very valuable as a software debugging tool. The digital troubleshooter needs to quickly verify whether a malfunction is a hardware or software fault. Some specific uses of the logic analyzer are program flow analysis, timing, and data qualification.

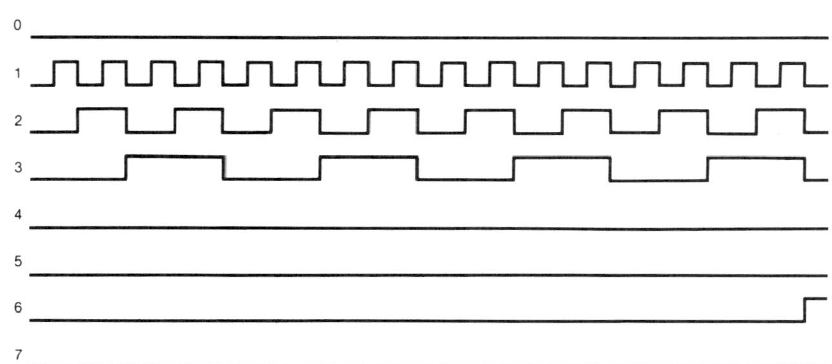


**Figure 7-17.** Timing diagram displaying the LSB stuck on 0.

Flow analysis involves tests that trace and verify program path. A subroutine can be traced and its path verified through various loops, jumps, or branches; or the main program path may be monitored to verify successive addresses.

Timing tests are those that monitor the duration of program execution. This might include the timing of an interrupt routine, an equipment start-up routine, or a program loop cycle.

Data qualification is where specific items of interest are captured and analyzed, and items not of interest are discarded: for example, specifying that only READ and WRITE cycles are to be captured and analyzed.

## Teleservicing

Using the logic analyzer as a remote diagnostic tool is one example of teleservicing. The logic analyzer of Figure 7-10 provides master/slave support so that one logic analyzer (designated master) can control remote logic analyzers (designated slaves) over telephone lines. This saves time for service engineers and technicians by allowing them to monitor and debug microprocessor systems remotely rather than on site.

With the auto-dial feature, a slave logic analyzer can be set up to automatically call the master when an intermittent problem occurs that it was previously set up to trigger on. To perform remote monitoring, diagnostic troubleshooting, and analysis, the system would be set up similar to that shown in Figure 7-18. To perform remote testing and repair of equipment, service personnel must learn the operation of remote test equipment and to think of more than just hands-on repair.

# A New Logic Analyzer: The Tektronix 1205

The evolution from oscilloscope to logic analyzer had its beginning in the early 1970s. Since then, the complexity, flexibility, power, and cost of logic analyzers have risen drastically. Today, a high-powered logic analyzer may cost well over $100,000. Since 1984, there have been several major manufacturers offering logic analyzers to meet the competition in the $10,000–$30,000 range. In 1987, the trend of lower prices may be established as Tektronix introduces their 1205 model.

The Tektronix 1205 logic analyzer, shown in Figure 7-19, provides basic logic analysis functions for well under $3,000. This logic analyzer,

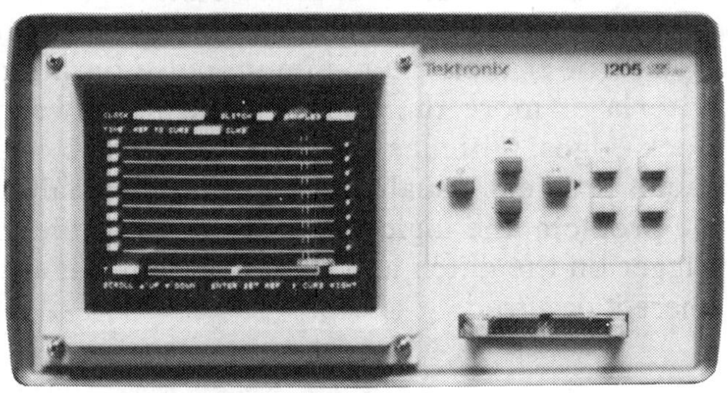


**Figure 7-19.** The Tektronix 1205 logic analyzer.

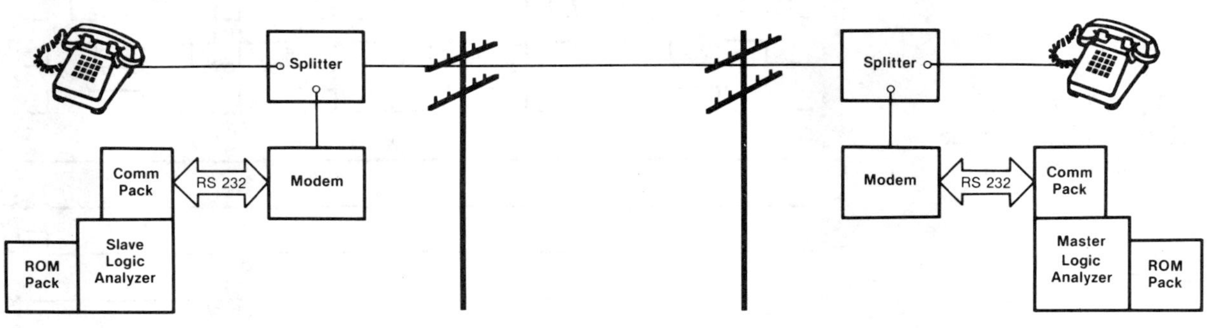


**Figure 7-18.** Teleservicing.

though small and much more affordable, supports state analysis, timing analysis, and microprocessor disassembly for some popular microprocessor analysis. It has a single probe connector that accepts specific probes for specific applications. Each probe automatically causes the 1205 to display the proper menu for that type of application. It also features glitch capture down to 10 ns on four channels for sampling rates of 25 MHz. Other Tektronic logic analyzers, such as models 1220 and 1230 offer many additional features at comparable low cost.

Traditionally, logic analyzer manufacturers have been eager to supply the maximum possible speed, word width, and memory depth. These features are for instruments tailored to the needs of design engineers rather than service technicians. Logic analyzers that collect data on more than 32 channels and at speeds approaching a gigahertz are expensive and have redundant capabilities. The technician's job is to troubleshoot and service the equipment, not redesign it.

The future of logic analyzers in troubleshooting digital and microprocessor equipment appears bright. The need for an instrument with which the technician can see the actual operation of digital circuits is essential. Logic analyzers provide an open window to this digital world.

# 8

# An In-Circuit Emulator

In this chapter, we break troubleshooting down to two major categories: (1) equipment and software that has worked for some time and has failed; and (2) equipment being developed or designed (prototypes) that has to be debugged. Emulation has become a standard technique for evaluating and debugging both hardware and software. This chapter discusses the use of the emulator in the development phase. We explain why the fastest way to get a prototype debugged and operating is to take one small part at a time, since problems tend to interact.

## What Is an Emulator?

In-circuit emulation (ICE) was originally introduced by Intel on its microprocessor development system (MDS). Intel has been a leader in supporting widespread use of ICE. Today, virtually every leading microprocessor manufacturer, as well as several independent development systems, have an in-circuit emulator (ICE) available.

### Two Categories of Troubleshooting

Troubleshooting, in general, can be divided into two categories: failed equipment and development of new equipment. No matter which category of troubleshooting the technician is faced with in digital and microprocessor work, there are related subjects with which he or she must become familar.

#### Failed Equipment

The first category is troubleshooting circuits that have been working correctly for some time but have now malfunctioned. Most books on troubleshooting deal exclusively with this category; and by far, the majority of our concentration in this book has been on this category as well. However, the second category does exist, and in this chapter we look at the development area of the digital and microprocessor world.

#### Development of New Equipment

Development of new equipment involves troubleshooting circuits that have never worked or circuits that have never been tried. This type of troubleshooting applies to prototype circuits in a developmental environment. It applies to circuits that are being produced on an assembly line or individually, as well as to an individual project. The development of any new microprocessor-based equipment must proceed as an integrated design sequence of steps along the software and hardware paths. Let's summarize the steps of each of these paths, beginning with software.

*Software Equipment*—If the software is simple, it may well be developed and written by one person. However, if it is complex, it is usually planned and partitioned into subprograms so that several persons or groups can work on it simultaneously. Careful planning and logical division makes the software easier to debug, link together, and modify. A detailed flowchart for each programming module is laid out. That flowchart is then translated into code. The code at this stage is usually a higher-level code, such as C or assembly. However, it must eventually be converted into machine code by an assembler or compiler. With the aid of several software development techniques and tools, each programming module is debugged. If logic errors are detected, the source code is modified, reassembled, and tested again. This may have to be repeated many times. The debugged modules are put together to form the complete software package. This package as a whole is tested and debugged.

*Hardware Equipment*—In the development of the hardware part of the system, the first step is the creation of a detailed logic diagram for each module. The important control signals should be described in terms of timing diagrams. Next, the components that will compose the system need to be selected to meet the requirements of such parameters as speed, power dissipation, and cost. After the components are selected, each module is constructed on a breadboard or wire-wrapped prototype board. These circuits are tested and compared to the logic and timing diagrams by tracing signal flow. It may take some extensive troubleshooting to correct both design and wiring errors. When these individual modules are working, they are assembled into a complete system called a *prototype*. The prototype is tested for wiring and other hardware errors before and after the software has been installed.

## Simulation

Often the terms *simulation* and *emulation* are erroneously interchanged. The simulation of a microprocessor is the process of executing a program written in the instruction set of the given microprocessor on a different computer, even a mainframe. That is, it refers to the functional replacement of a hardware device (the microprocessor chip itself) by a program. So it is said that the device is simulated by software. The program will respond to the same inputs as the microprocessor and produce the same outputs. The biggest disadvantage is that the simulation process is much slower than the actual device running at real time.

To simulate the architecture of the microprocessor, a block of memory is set aside in the host computer. The registers, program counter, and accumulator are represented by memory locations, and each instruction of the simulated microprocessor is achieved by several instructions of the host computer. This is the main reason that simulation is much slower than real time. The contents of these memory locations are the same as the register contents would be if the same program were executed on the microprocessor being simulated.

In addition to the microprocessor, simulation is used for ROM simulation by executing programs out of RAM as if they were in ROM. An initial program, during its development, will contain a number of bugs and should not be placed in final ROM or PROM until it is free of problems in logic. Remember that a simulation cannot detect time-dependent errors, such as those associated with processing interrupts and interfacing with certain timing-critical I/O devices. Simulation of the ROM is much less complex than simulating the microprocessor. Simulating the microprocessor is used when either the actual device is not available or for convenience when debugging a program. When the timing must be checked in the development of a system, or the input and output must be checked in real time, the most useful and significant aid is the emulation of the microprocessor. This is done with an ICE.

## Emulation

The process of emulation is essentially simulation performed in real time. An emulator, therefore, is a real-time debugging aid consisting of some hardware, software, and an MDS or a host computer for the ICE hardware and software. With emulation, the software can be tried out in the target system hardware while retaining the facilities of the development system to monitor program execution and to change the program if necessary.

The ICE facility is provided by a special PCB inserted in, or otherwise connected to, the MDS or host computer. Emulation is usually controlled by a structured, high-level language in which sophisticated test procedures can be written. The system under development is connected to the ICE by a special cable called an *umbilical cord*. This ICE facility is shown in Figure 8-1. The microprocessor has been removed from its socket on the prototype breadboard, and the emulator connected in its place with a 40-pin connector on the end of the umbilical cord 40-line cable. The difference is that all the signals carried by this cable are generated by, or under the control of, the ICE rather than by the actual microprocessor.

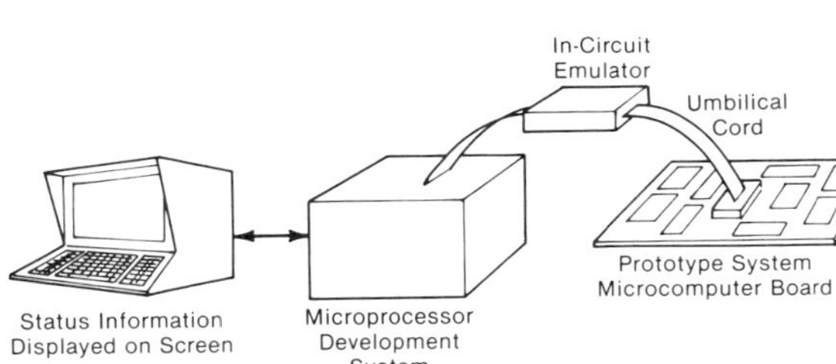


**Figure 8-1.** An in-circuit emulator facility.

## Debugging

By using the software and hardware of the ICE to emulate the microprocessor, we can completely control and test the prototype of the system under development. With this facility, it is possible to stop the operation of the microprocessor and examine, or even change, the registers. The troubleshooting advantage of this is clear, since the contents of the registers of a microprocessor are not available. Only the values on the bus are available.

Using the ICE, we can stop the operation of the microprocessor automatically, using breakpoints in the program. The troubleshooter can, from the keyboard, execute actual I/O instructions, such as operating peripheral equipment. The processor can be stopped repeatedly to examine the bus levels, registers, and memory locations. All of these operations can be performed in conjunction with the other aids, including those for software development in the MDS. Examining or changing the register or memory content using symbolic form rather than binary (machine) form is called *symbolic debugging*—a timesaving and frustration-relieving feature.

When the ICE facility is using breakpoints to specify where the program will automatically stop, an additional feature called *trace-back* can be utilized. This capability provides what is commonly called a "snapshot" of the history of the signals on the bus for a specified number of machine cycles. Whenever an error is encountered at a breakpoint, it is not usually caused by the instruction at the breakpoint but by some earlier instruction in the program. The debug problem is to locate the particular instruction that caused the error. The trace-back approach of examining the previous instructions can be used to determine the exact cause. This process can be repeated until the location of the instruction causing the problem or even the mistiming is located and corrected.

The ICE is certainly a debugging tool. It provides the first opportunity for checking the operation of the complete system, including I/O ports and interfaces. Most of the MDS's software and hardware resources can be shared with the prototype system during the debugging stage. As we have seen, a program can be executed step by step in the prototype while the status is being displayed, just as if it were being run in the MDS. Additionally, hardware items on the prototype, such as I/O interfaces and memory devices, can be tested and debugged in the development system's memory. Once the prototype is completely debugged, the ICE can be removed from the prototype and replaced by the actual microprocessor.

# Examples of Emulators

Several companies make emulators. Some are designed to work only with certain MDS. Others work with the IBM PC and other compatible computers. The hardware of an emulator consists of control circuitry, memory to store the trace data after each instruction executes, and an umbilical cable with a plug at the end. It contains a microprocessor that will run the test program under control of the emulator, then give a window into the operation of the circuitry on the prototype under the control of the MDS or PC. Some emulators have an additional pod, such as those used on logic analyzers, so that a trace of the sequence of the hardware signals on a bus can be captured to check timing.

The software of the emulator is similar to a powerful monitor program or debugger program as discussed in Chapter 5. The emulator commands and the MDS memory can be used to test each part of the prototype.

After this brief and general description of an emulator, we now look at some specific features of some examples of emulators.

## Intel ICE-86A

The Intel ICE-86A provides hardware and software debugging capabilities for 8086 microprocessor systems. It also includes ICE extensions to debug systems that include peripherals such as the 8089 I/O processor and the 8087 numerical processor extension. This ICE includes three circuit boards that reside in any of Intel's Intellec MDS. The umbilical cable and buffer box connect the Intellec system to the prototype by replacing its 8086 microprocessor. Using the ICE-86A module, the 8086 software can be executed in continuous or single-step modes. The Intellec system's memory can be substituted for the prototype's memory and breakpoints allow the program to be stopped at specified conditions. The trace capability gives a detailed history of the program execution prior to the breakpoint.

The ICE-86A emulator allows for the various stages of software and hardware development to proceed interactively. This is more effective than independent software and hardware development followed by system integration, which has been the traditional development method. With the ICE-86A module, prototype hardware such as peripherals and interfaces can be added during the design of the total system. Software and hardware can be tested in stages while the total system is being designed and developed. A block diagram of the ICE-86A is shown in Figure 8-2.

Access to the prototype system software can be done symbolically by referring to the source program variables and labels. Symbols can be substituted for numeric values in any of the ICE-86A emulator commands. This eliminates the need to remember the addresses of variables, program subroutines, I/O ports, or data. The use of symbols for these statements and procedures simplifies greatly both tracing and breakpoint setting.

Intel makes several ICEs for different microprocessor systems, including the ICE-88 for the 8088, ICE-22 for the 8022, and the ICE-80 for the 8080 microprocessor.

## Hewlett-Packard HP64000

The HP64000 logic development system offers a wide selection of emulators to support various

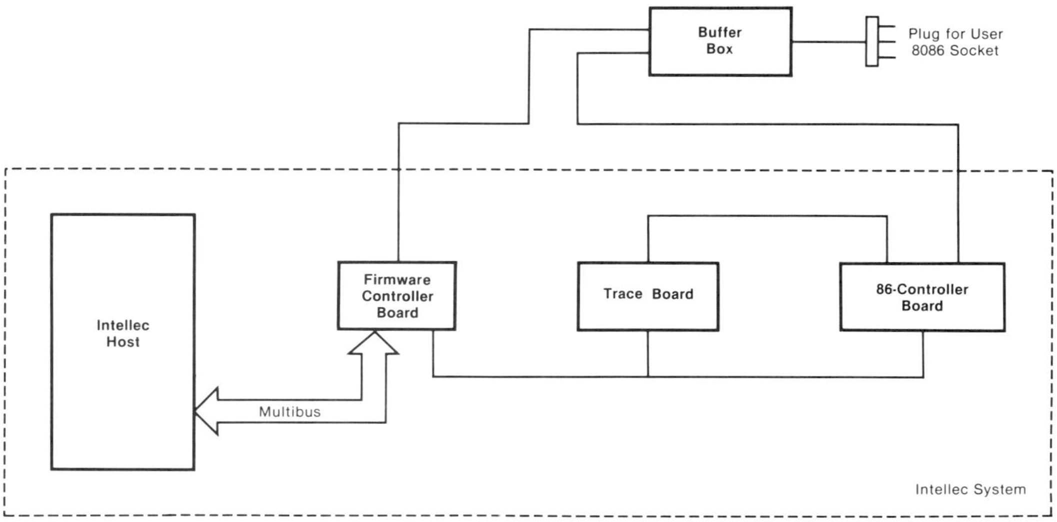


**Figure 8-2.** Block diagram of the ICE-86A Emulator. *(Courtesy Intel Corp.)*

microprocessor-based product development. The emulator programs developed on the HP64000 system, or on compatible computers, are run on the emulator subsystem for real-time debug and analysis of the prototype. This emulator also allows single-step, display, and modification of memory as the microprocessor is run. Some of the features of the HP64000 emulation subsystem include

- Real-time emulation mode without inserted wait states
- Run-time controls for single cycling and register display
- Mapping memory blocks to emulation or prototype (target memory)
- Simulated I/O using the HP64000 system resources
- High-speed emulation memory

The HP64000 system currently supports over 30 microprocessors manufactured by eight different companies. For microprocessors not presently supported by a dedicated system emulator, a custom emulator can be developed using their user-definable emulator as a base.

An emulation bus analyzer should be added to the system to monitor activity on the emulator bus. With it, real-time traces of address, data, and status control signals can be displayed in the microprocessor's mnemonics or in an appropriate numerical base.

The HP64000 offers five analysis subsystems to be used for troubleshooting, debugging, and optimizing the target system software. These analyzers are microprocessor specific and usable on programs written in high-level languages, such as Pascal or C.

The HP64000 system supports microprocessor product development through all phases of design, development, production, testing, and modification. It offers full microprocessor emulation as well as several other features for debugging, integrating, and optimizing by providing real-time nonintrusive measurements to quickly locate problems.

## Tektronix V-System

The development system by Tektronix for both 8-bit and 16-bit microprocessors is their V-System. This system is configured to run on several computers, including the IBM PC AT/XT. In these systems, as in other emulators, real-time emulation is accomplished by executing code on a microprocessor identical in function to the one targeted for the prototype hardware. The emulator processor of the V-System is run under the control of powerful debug software that allows both control and tracing of the code's execution.

The trigger trace analyzer (TTA), an option to the Tektronix emulator, uses a high-speed trace buffer to capture real-time software and hardware logic events with the prototype running at the design's full specified operating speed. Several word recognizers allow definition of sophisticated triggers when tracing code execution.

The V-System enables you to incrementally integrate your prototype software and hardware. As shown in Figure 8-3, starting with mode 0, you can do a majority of your software design without a prototype; that is, mode 0 debug can begin long before actual hardware is available. Once prototype hardware is available, you can gradually move from using emulation resources exclusively to using prototype resources, as shown in Figure 8-4. The final step is to use all of the prototype resources to execute the program. Tektronix calls this "mode 2." The emulator in mode 2 is said to be transparent, which means

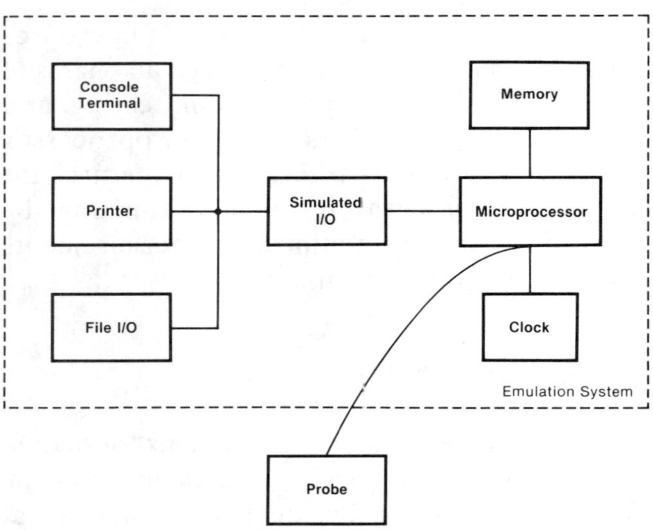


**Figure 8-3.** Modes of emulation: phased emulation, mode 0.

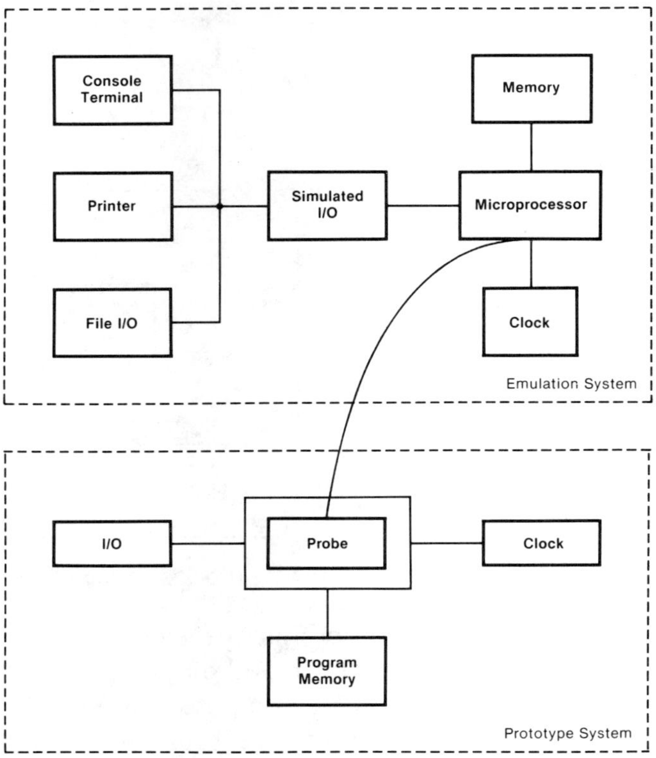


**Figure 8-4.** Phased emulation, mode 1 and mode 2.

that the prototype will perform as it will in actual operation with the emulator disconnected. With this feature, one need not spend time working around the development system.

As stated at the beginning of this chapter, we can troubleshoot equipment that has failed after working for some time or equipment still being developed. Most troubleshooting falls into the first category. However, if you are in a position to do troubleshooting or debugging of developmental prototypes, you will see that the development cycle will be at least slightly different for each instrument. The main points in using emulation are to develop and test both the hardware and software together in small modules. In order to speed up the debugging process, it is important that you take the time to use as many features of the MDS and emulator as possible.

# 9

# Signature Analysis

To maximize efficiency in the task of digital troubleshooting, manufacturers are continually developing new types of test equipment. Foremost among these is the signature analyzer. This chapter shows its basic operating principle and how it provides a simple way to quickly check digital circuits by comparing the digital signals they produce. Examples are shown of finding faults by using signature analysis to test digital circuits built in production quantities. Signature analysis is presented as a troubleshooting technique that locates faults to the component level in microprocessor-based circuits. Signature analysis is extremely accurate and repeatable, because there is but one signature per node. Signature maps are also introduced.

## Basic Operating Principles

Signature analysis can be a fast and accurate troubleshooting method for microprocessor and other digital systems. It is a way of uniquely characterizing the proper operation of a circuit. Finding malfunctions is reduced to tracing signal flow and comparing measured signatures to those previously recorded. They may have been recorded either on a troubleshooting tree or flow diagram, or in a reference memory of the instrument or an interfaced computer-aided system.

A data stream normally has a unique binary bit pattern. This pattern is called its *signature*. If this pattern is known, the technician can determine whether a circuit or system is functioning properly by comparing the measured signature with a predetermined signature. If there is a match, then the technician moves to the next circuit node or test point. If there is no match, the faulty component must be located and replaced, and the signature must be rechecked. Thus, each node in a circuit or system has its own unique signature when it is working properly and under the same test conditions. A number of signatures can be obtained from a given digital or microprocessor system, but a repeatable data stream must be selected on a correctly operating system. Once a signature of a data stream is defined and stored or recorded, only a correctly operating system is capable of producing the same signature. Therefore, even difficult-to-detect data errors on bus-oriented digital systems can be reliably detected. This test method is dynamic rather than static, because it makes its signature analysis under its normal operating conditions.

Actually, the signature analysis technique is based on the traditional method of signal tracing. In analog troubleshooting, a technician follows an annotated schematic that depicts the correct waveforms and voltages at various points in the circuit, which can be observed with an oscilloscope and voltmeter. An annotated schematic of part of a television receiver is shown in Figure 9-1. To do this with long, high speed digital data bit streams would be extremely frustrating, perhaps impossible. Even if there were room on the schematic to annotate a data stream, it could not be compared with what was seen on an oscilloscope or a logic analyzer except in the simplest and most special case. If only one small data bit was missed in a digital data system, and there can be literally millions of bits to examine, the pattern would be useless as a fault indicator. Here the signature analyzer comes to the rescue. It takes a complex data stream from a test node on the PCB and converts it into an easily interpretable 4-digit hexadecimal signature on an LCD or LED display. This test method is also unique because it is dynamic; that is, it runs at the system's full operating mode and speed. These 4 digits are perfect representations of the data streams. They are easy to annotate on schematics, which makes them ideal references to use for comparison against signatures obtained by the troubleshooter as the system is being tested.

So, the signature alone means nothing, but the fact that this signature must always be present at the same test point of every identical, properly operating digital system means everything.

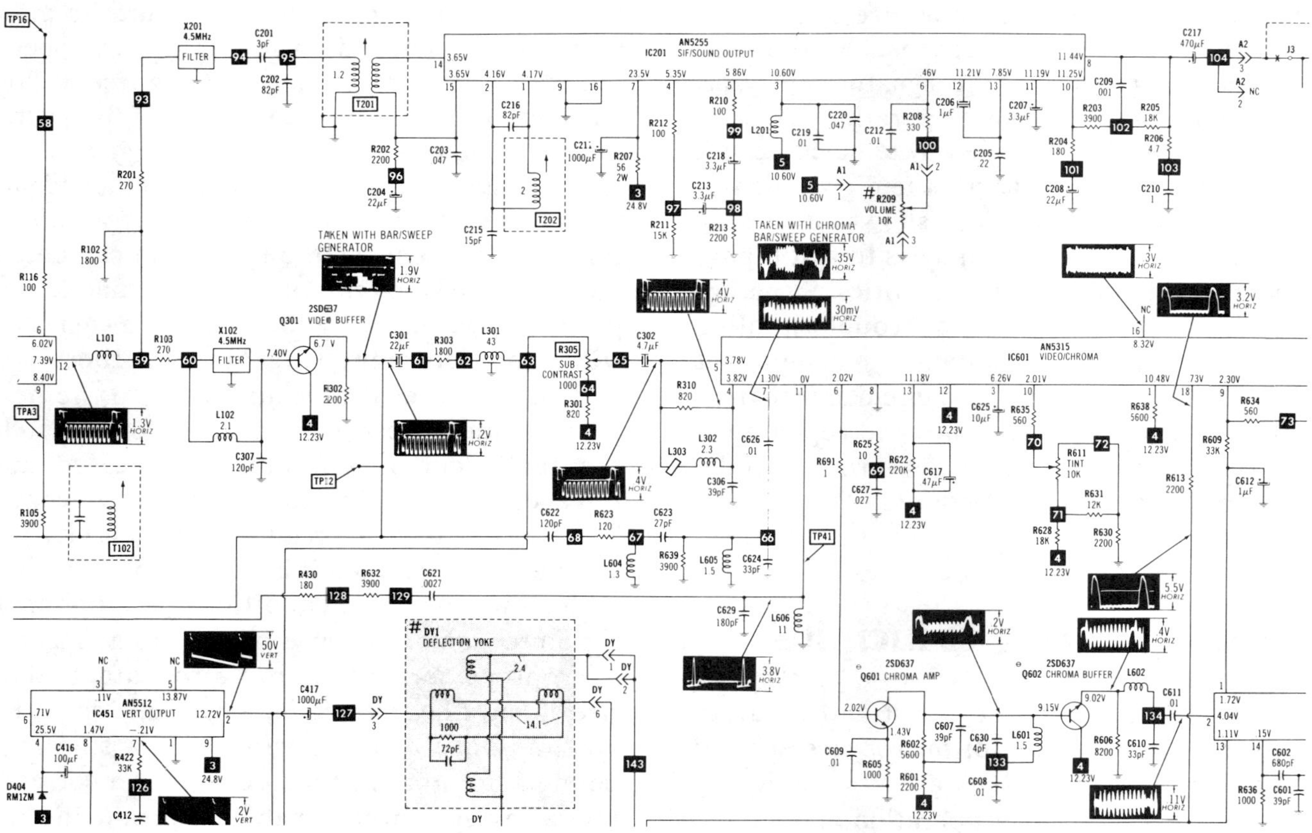


**Figure 9-1.** Schematic of a portion of a television receiver. *(Courtesy Howard W. Sams & Co.)*

## Terms

Some terms will probably be new to the technician first encountering the signature analyzer as a troubleshooting tool. Some terms are actually part of a unique and specialized vocabulary associated with signature analysis. Let's begin with stimulus. *Stimulus* is the agent (active force or real information that produces a result) that influences the circuit to produce the data stream. The stimulus can be a program in RAM that exercises parts of the digital system and creates the data stream. The data stream produces a waveform as a result of the stimulus, which, in turn, is applied to the input of the signature analysis.

One of the most useful and fundamental ways to create a data stream that checks a microprocessor is to program the microprocessor in the NO OP mode as a simple 16-bit counter. These serial pulse trains on the microprocessor's address bus lines can be easily checked with a signature analyzer. Since each line has a different pulse train representing a data stream, the signature on each line is different. The signatures will, however, be consistent when similar microprocessor circuits are tested. This procedure is called *testing the kernel*, which is the basic signature analysis test of a microprocessor chip. Because this test checks a large part of the microprocessor itself, it is an efficient procedure. It is only necessary that the power supply be correct and the interrupt lines be in their proper logic states to perform the kernel test.

Sometimes the data stream is thousands or millions of bits. It is sampled only during a certain time period, called the *measurement window*, which depends on its speed or data rate. The signature is then two or four hexadecimal digits generated from the data stream during the measurement window. This manipulation of a long data stream into a signature is called *data compression*. The point in a circuit where the signature analyzer samples the data stream is called the *data node* or the *V-node*.

## Where Can the Signature Analyzer Be Used?

Signature analysis was introduced in 1977 by Hewlett-Packard as a digital troubleshooting technique. It was a significant advancement in digital troubleshooting, but the service industry has been slow to accept the technique. In the class of instruments in which the signature analyzer is a part, few actually specialize in troubleshooting. Most are intended primarily for use as development tools, such as the in-circuit emulator and logic analyzer. The signature analyzer, on the other hand, is especially efficient as a troubleshooting tool but has little value as a development instrument.

There are several reasons for the lack of widespread acceptance of signature analysis. On the surface, it appears to be complicated and confusing to use, so many technicians reject it. They feel that they do not have enough time to initially set up a reliable signature analysis procedure. They fail to realize how the time spent will be more than recaptured in the time saved in future troubleshooting. This is especially true if the technician works on a limited number of different products, because less initial documentation is needed to get the value of the time saved.

Where the signature analyzer is used depends greatly on the efforts of product manufacturers. If the product is designed with signature analysis in mind, the service literature will contain documentation with flowcharts and signatures to allow the technician to systematically troubleshoot it by probing test points and looking for correct signatures. The first device with incorrect signatures on its output is the faulty component or unit, and it should be replaced.

In a company with enough volume of PCB repair to justify being equipped with expensive automatic test equipment (ATE), the signature analyzer can be used to free valuable ATE time. PCBs often get unnecessarily recycled through the ATE system when the product failure is such that the ATE system is not equipped to pick up some particular faults. Often the signature analyzer can be used to test for timing, interface, or interconnect problems, and certainly intermittent problems, that the ATE system is not designed to find. This can be done with the boards being driven at full operating speed.

It may still seem that signature analysis is not for every troubleshooter, at least until more product manufacturers adopt it as a service standard. However, if a technician can implement even limited signature analysis into his or her troubleshooting techniques, it could very well be invaluable.

### Documentation

When documenting good signatures, several points should be considered. First, enough points (nodes) in the hardware must be chosen so that the problem can be isolated to the desired level. More points must be used to isolate a problem to the component level than to the board level. The number of nodes documented might be slighted, because for each point where a signature is acquired, a clock source and edge must be specified. Also, a source and edge must be given for a start signal and a stop signal. The data between these two signals will form the signature.

In addition to the clock source and edge, the operating mode of the microprocessor system being tested must be specified in order to have consistent signatures. For example, the system diagnostic mode might be to sequence through blocks of addresses.

A list of symptoms and possible causes can be helpful, especially if they are used with microprocessors that can display error codes when a specific failure occurs. Also of value is a troubleshooting tree, which involves taking a signature and showing a branch based on the results of the test. A good troubleshooting tree requires much more careful development than any other aid or approach, because a troubleshooter following a fault through the tree could get lost in a wrong branch. By following through a set of special signature lists, the technician can use a troubleshooting tree to trace systematically through a circuit.

A technique called "growing the kernel" is used to produce valuable documentation. Here the instrument is exercised in a series of operating modes that are arranged in increasing levels of complexity. In this way, each successive operating mode exercises a larger percentage of the microprocessor system. By observing the first operating mode in the sequence that fails, the technician can quickly limit the problem to those assemblies or units that are used in the failed operating mode but are not used in the previous modes that all passed.

A signature map is also a method of documenting equipment for signature analysis in troubleshooting. A signature map gives the technician both the signature and the location of the node at the same time. A signature is printed directly on a PCB next to the test point where the signature is to be taken.

## How a Signature Analyzer Works

A *signature* is a unique abbreviation of serialized binary numbers that occur in microprocessor control and computer systems. It can be thought of as a fingerprint that identifies serial binary data bit streams. Just as a person's fingerprint is only a small part of the entire person, though it contains unique patterns that accurately identify the entire person, so a signature of only 16 bits in length can identify a data stream of millions of bits. Although the signature size is only a small part of the entire data stream, the signature nonetheless contains unique patterns that accurately identify the entire data system.

### A Basic Premise

Signature analysis is based on the premise that if a known test routine is operating in correctly performing hardware, the pattern of pulses at each circuit node in that hardware can be characterized. If that same routine is applied to faulty hardware, the pattern of pulses at the circuit node will be different in ways that can be observed by instrumentation. A signature analyzer treats the binary data present at a circuit test point as a number. This number must be uniquely generated by a known and specified test routine. A problem that immediately arises is that this sort of binary number would be too long to document easily or to display in a form easily read.

The solution to this problem comes to us out of one of the error-detecting schemes used in mass storage and transmission of data—cyclic redundancy checking (CRC). This kind of error detection is done by applying a mathematical operation to the data to reduce extremely long binary numbers to a 16-bit binary remainder of a long-division operation. See Figure 9-2 for an idea of how the signature is mathematically obtained from the data. This remainder is then converted to a special kind of hexadecimal notation for easy reading and documentation. This special character set (0, 1, 2, 3, 4, 5, 6, 7, 8, 9, A, C, F, H, P, U) was chosen for easy interpretation on a 7-segment display. In practice, this is done with a feedback shift register (FSR) with

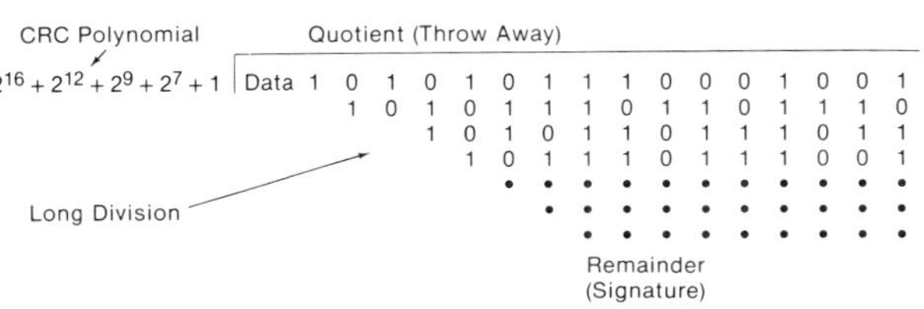


**Figure 9-2.** Division of data stream by CRC polynomial.

feedback loop tapes in certain defined places that do the best job mathematically of finding errors. As the data bit stream of any length is accepted through the data probe, it is marched through the 16-bit linear FSR in the analyzer, as shown in Figure 9-3. It is clocked into the input of the register from a defined start to a defined stop point. At the stop point, the content of the register is the remainder of the division. This number is uniquely determined by the data. Odds are 1 in 65,536 of different data producing the same number. This feedback portion of the shift register allows for just one faulty bit in a digital bit stream to create an entirely different signature than would be expected at a properly operating test point.

## Overview of a Block Diagram

A signature analyzer model that is popular in microprocessor system troubleshooting is shown in Figure 9-4. It is a test instrument for troubleshooting complex electronic logic circuits, especially microprocessor based, to the component level. Its block diagram is shown in Figure 9-5, and a brief description follows.

The central control unit (CCU) uses a microcomputer, which contains RAM and ROM. This CCU monitors and directs the overall instrument operation, controlled by the stored instructions in ROM. The keyboard matrix is scanned by the microcomputer for any pressed keys, which then responds by setting control lines or changing the measurement.

The signals are input to the signature analyzer through the data probe and timing pod, which are terminated in the input voltage

**Figure 9-3.** Shift register portion of the signature analyzation.

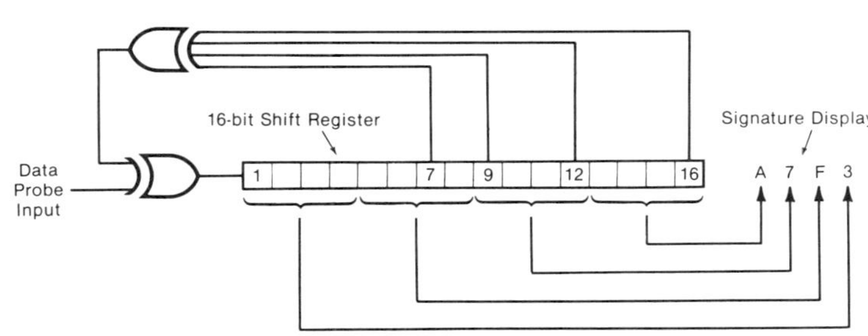

comparators and level shifters. The input signals are compared with reference levels and are level shifted to TTL levels. They are then selectively passed or inverted by the edge-select and qualify enable circuits, depending on the chosen polarities and the signature analyzer mode.

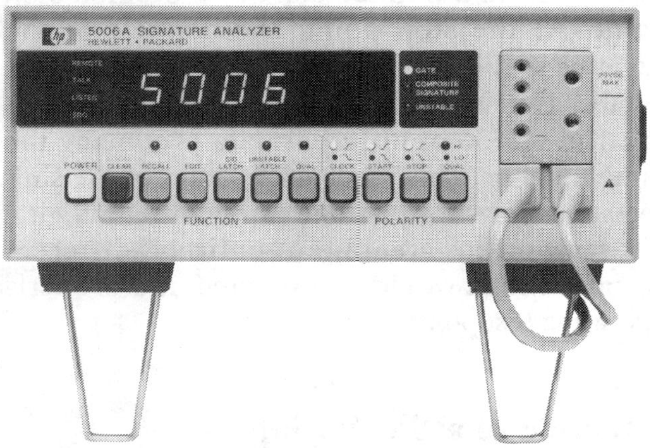


**Figure 9-4.** Hewlett-Packard 5006A signature analyzer. *(Courtesy Hewlett-Packard)*

The gate control circuit is a free-running synchronous state gate driven by the timing signals from the timing pod. This establishes a window around the data of interest. The gate handshake circuit is used to interface the CCU to the measurement window, which helps the microcomputer determine when to enable measurements.

The data sampler circuit examines the incoming data once each clock edge. This synchronous data stream output is then routed to the FSR. This FSR is the 16-bit serial shift register mentioned in the previous section and shown in Figure 9-3. The 16-bit residue (remainder) from the FSR is formatted by the microcomputer to produce the four-character signature of the original data stream. This four-character memory is then output through the display driver circuits to the front-panel display.

## The Probe and Pod

The Hewlett-Packard model 5006A signature analyzer has a probe with dual threshold levels that are compatible with TTL logic. It also has a lamp that serves as a TTL logic probe that is similar to the one discussed in Chapter 6. In this capacity, it provides additional troubleshooting information by indicating a logic high or low, a bad level, or pulsing states.

The circuit under test is also connected to the signature analyzer with three gating signals: start, stop, and clock. The start signal begins the measurement window, which prepares the shift register to receive information from the data probe. The stop signal closes the measurement window. Then the clock samples the system clock of the device under test to ensure the synchronous acceptance of data and gating signals by the shift register. The front panel has pushbutton control of the particular active edge (positive or negative slope) of these gating signals, which gives the user more latitude in troubleshooting. The probe and pod are shown in Figure 9-6.

## Input Threshold and Gating

The Hewlett-Packard 5006A is preprogrammed to trigger on the standard TTL logic thresholds. A two-level threshold is used, as described with the logic analyzer, for the data probe input: 2.00 V and above for a high and 0.80 V and below for a low. The other inputs—clock, start, stop, and qual—all use a single-value TTL mean of 1.40 V.

So the 5006A signature analyzer is preprogrammed to trigger on standard TTL logic levels, though the actual trigger threshold may be set to any level between +0.9 V and +14.0 V to accommodate CMOS logic. To set the 5006A threshold for other than TTL levels, simply connect the front-panel connector labeled CMOS to the equipment power supply. Automatically the DC voltage value is sensed and the program threshold levels are established.

The front panel has a gate light that indicates proper start/stop gating operation. It remains on during the measurement window, but also has a stretching feature to make it visible during very short gating times. If the signatures

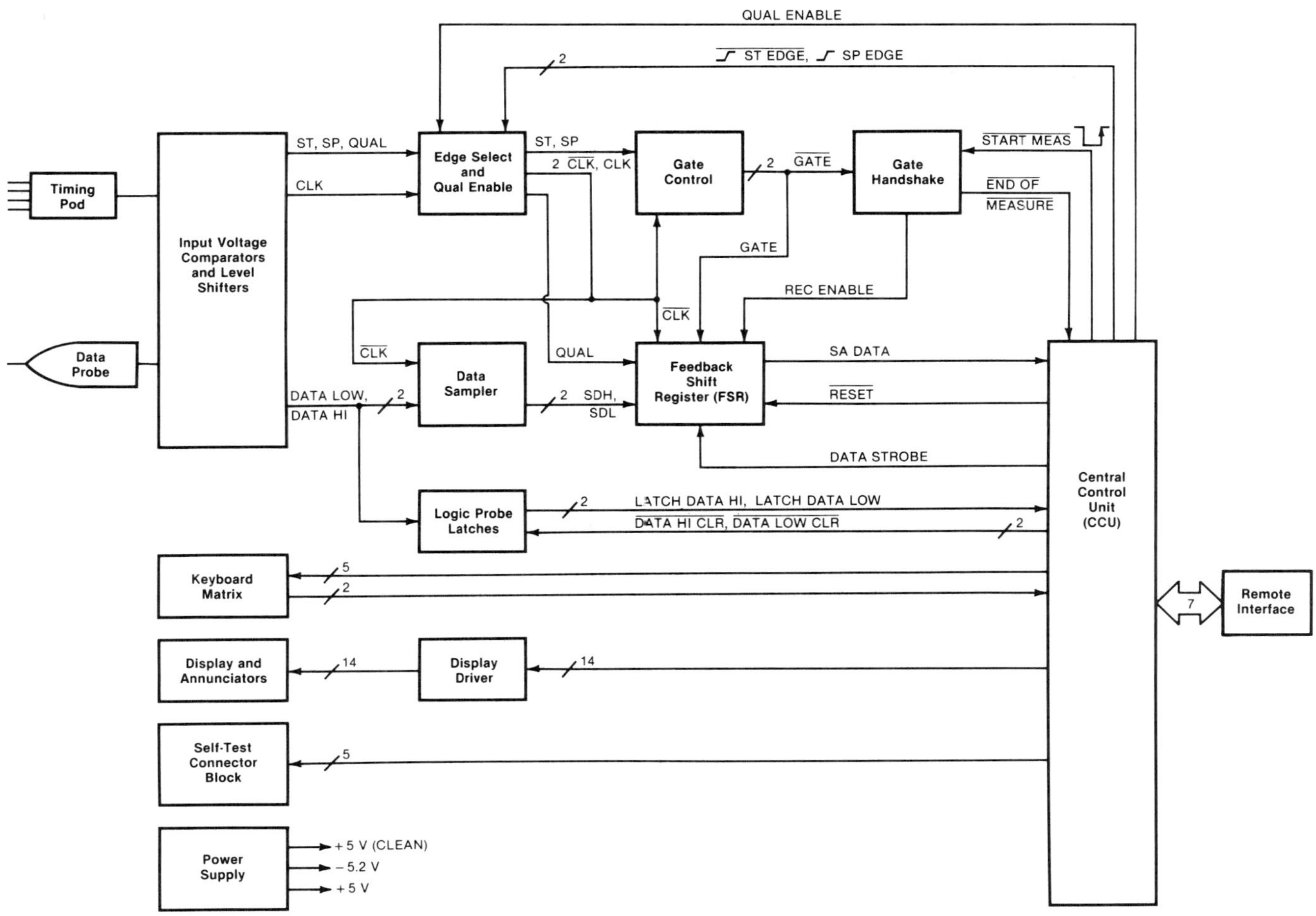


**Figure 9-5.** Hewlett-Packard 5006A block diagram. *(Courtesy Hewlett-Packard)*

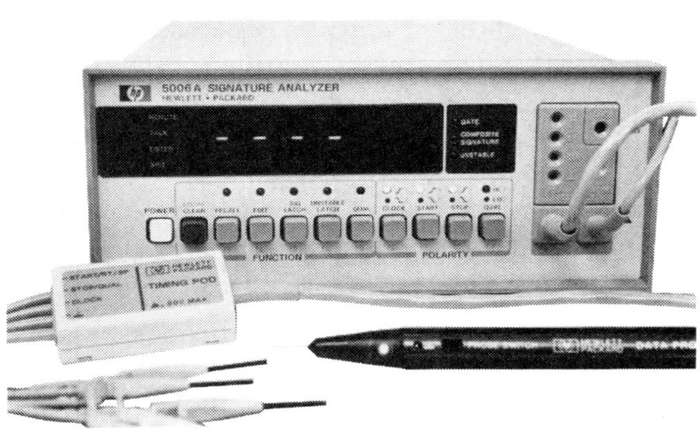


**Figure 9-6.** Probe and pod of a 5006A signature analyzer.

at a node are different in adjacent windows, the unstable signature light indicates this, which should alert the technician to intermittent faults that may not be apparent from the signature display. If the gate does not open or close as expected, it is probably because there is not a clock edge both before and after the start and stop edges.

## Using the Signature Analyzer in Troubleshooting

Although the evolution of the microprocessor has made designing easier in most electronic systems, it has also created a troubleshooting nightmare. Fortunately, the signature analyzer presents a solution in many cases.

## Modes of Operation

The signature analyzer is used in one of two modes: program-driven or free-running. The program mode uses a test routine stored in a section of ROM in the equipment under test. This program exercises the system and provides known signatures in the form of fixed sequential bits on the bus lines. In the free-running mode, an instruction such as NOP (no operation) is used to place a single byte or word on the microprocessor data bus, which has been isolated from the system data bus with a special plug fixture or with some other pin isolation technique. If the equipment has been designed with signature analysis troubleshooting features built in, it will have jumpers on the hardware for this purpose. The NOP instruction causes the microprocessor to act as if it has fetched a NOP at each sequential address in memory, so it increments the instruction address and performs another fetch. This continues until it has provided all possible address codes on the bus or until power is turned off for the system.

As the microprocessor cycles continuously through its entire address field, start/stop (gate) signals are derived from the address bus lines. To implement this free-run mode in most microprocessor-based systems, all that is required is the addition of some switches and pull-up resistors. With the switches installed, the data bus can be opened and the microprocessor forced into a free-run mode.

Free-run signature analysis has the advantage of not requiring any ROM space for storing the stimulus program. Software-driven signature analysis, on the other hand, has the advantage of being able to exercise a greater portion of the circuitry of the instrument under test. For thorough testing, both techniques could be implemented in the same instrument.

## Checking Out the Kernel

To check out the microprocessor section of a system, the microprocessor itself is forced into a free-run mode by opening the data bus lines and executing a NOP instruction. The kernel area constitutes the address decoding, the primary program ROM, and the microprocessor itself. This kernel area, once separated from the board system, can be easily checked for proper operation by running the stimulus program to aid in the rest of the testing. The use of address decoding is also important so that the troubleshooter can quickly and easily isolate ROM failures.

Provision should be made for proper initialization of all boards. The objective, through the use of sufficient reset circuitry, is to have the system power-up or reset to the same condition every time. This repeatable, predictable condition is a necessary requirement for signature analysis.

While running the signature analysis test on a 6800 microprocessor, verify the functioning of the kernel by taking a single signature. If the kernel is operating, place the signature analyzer's probe on the processor's +5-V supply to produce a bit stream of 1s, which results in a signature of 28PH. Figure 9-7 shows the result of a particular stimulus applied to a good 6800 microprocessor in a system. In this case, a good signature at pin 2 is enough to indicate that the processor kernel is operational. Take other signatures if this test or further testing indicates a problem. This signature test provides about an 80% confidence level that the microprocessor is good. Each pin of the 6800 can be probed while the test is run. The result will be the signatures in Figure 9-7 and a confidence level of better than 95%. The testing of each signature on the microprocessor pins takes less than a minute. After you verify the kernel's operation, check the address lines and decoders for proper operation.

In the free-run test, the microprocessor runs through its entire address repertoire. By placing the signature analyzer's data probe on

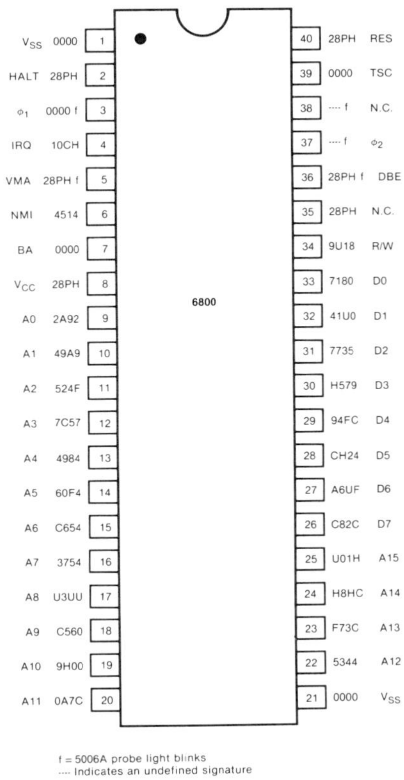


**Figure 9-7.** Running a particular test on a good 6800 microprocessor yields this set of signatures.

each of the address lines, you can obtain a signature for each one. These signatures will be the same for any 6800 microprocessor tested in this way.

The same test will provide signatures at the outputs of the address decoders. However, since their output depends on equipment design, the signatures will vary from one design to another.

After verifying the microprocessor's ability to access the various components of the system, you can check these components. The most logical area to investigate next is one closest to the microprocessor, such as the ROM in which the operating progam is contained.

## Checking Out Memory

Though the exact procedure depends on the design of the equipment under test, the ROM can be checked with a single signature or one signature for each ROM chip in it. By placing the data probe of the signature on a specific data line, you can take a signature that is unique to the content of the signature being tested.

Check the RAM using a diagnostic program to generate the data stream. A stimulus for the RAM tests can be loaded from the monitor ROM to RAM. The data transfer is accomplished by a short utility program located in the monitor ROM at some specific address. In this case, you would need signatures for the tests in the form of tables or schematics of the system under test when that system was known to be in proper operating condition.

## The Clock and Timing

The clock and timing circuitry is self-stimulating; that is, it runs independently of the microprocessor and any program associated with it. It is therefore also asynchronous to the system. For this reason, the start, stop, and clock inputs to the signature analyzer are taken directly from the part of the circuit being tested. This means that the signature tests are carried out at the clock rate of the circuit, which is very desirable, because taking signatures in this way is simpler and the results are more easily interpreted.

## Loops and Feedback

When switches, pull-up resistors, or jumpers are being designed into the system for signature analysis troubleshooting capabilities or being retrofitted into existing equipment, another consideration is to allow the opening of all feedback paths and loops in the system so that a bad node does not propagate bad signatures throughout the system.

Along with, and closely akin to, feedback loops and the confusion they can present is that any three-state device can slow the settling of a logic level. This can cause a mistaken signature.

To minimize this, the three-state devices should also have their own pull-up resistors.

The bottom line is that signature analysis is not for everyone, at least until more product manufacturers follow the lead of Hewlett-Packard and adopt it as a standard troubleshooting technique. However, if you can implement even limited signature analysis into your troubleshooting program, you may find it invaluable. It offers a cost-effective test method that can efficiently be used to troubleshoot complex digital and microprocessor systems.

# 10

# Repair, Checkout, and Preventive Maintenance

Even after the trouble is found and the faulty component is located and replaced, the troubleshooting effort is not necessarily completed. An operational check will verify that the equipment is free of all faults and performing properly again. Several helpful hints are provided here, including documentation, checking for intermittent problems, running appropriate diagnostic programs in a loop while the computer is being checked out, and checking not only the original trouble symptoms but being aware of entirely different sets of symptoms. Precautions, such as things to be careful of when you reassemble the equipment so that you don't insert a new problem, are included. Techniques for preventing damage are covered in the discussion on digital system hazards in the workplace.

## Eliminating the Trouble

Once the trouble is located, it has to be eliminated or cleared. Depending on company policy, or other factors in a particular situation, it may not be the troubleshooter who actually clears the trouble. For instance, electronic manufacturing companies often have their troubleshooters locate the problem in a piece of equipment that fails the testing procedure of the assembly line. The troubleshooter fills out a tag instructing an assembler to replace a defective IC or capacitor. The equipment is then sent back to the assembly line test, where, hopefully it passes inspection. The troubleshooter rarely knows for sure if the trouble was cleared unless he or she gets back that particular piece of equipment to troubleshoot again.

The maintenance person at a manufacturing plant faced with a problem in a programmable controller occasionally must resort to calling in an "expert"—a representative from the controller manufacturer. The expert arrives on the scene, troubleshoots the equipment, then turns over the elimination of the problem as well as preventive maintenance to the maintenance person.

Eliminating a trouble is different from locating a trouble and thus requires a different set of skills. The quality of workmanship of the repair separates the professional from the amateur technician and reflects not only on the individual but on the company as well. Obviously, it is important to do a professional job. In the first part of this chapter, we discuss some skills required to properly eliminate a problem from equipment once that problem has been located.

### Some Basic Considerations

First, to protect the equipment and yourself, disconnect the system from the power line before removing a component, clearing a short, or attempting a mechanical procedure. Any signals from test equipment, such as pulse generators or pulser probes, should also be removed from the equipment before attempting to clear the trouble. This is particularly important when working with CMOS ICs and FETs.

## Replacement of PCB Mounted Components

### Hand Tools for the Digital Technician

Some specialized tools are required for digital trouble elimination, in addition to the common hand tools used in electronic equipment maintenance (such as long-nose pliers, sidecutting diagonal pliers, screwdrivers, nut drivers). These special tools make the life of the digital technician much nicer. They will be described throughout this chapter.

Because of the miniaturization of components and equipment, a set of small tools is a must. A small pair of wire cutters is important. Its jaws should meet all the way to the tip, and one side should be flat, without a sharp bevel, to allow precise trimming of component leads with a flat cut rather than a sharp tapered end.

Several types of small needle-nose pliers with various styles of blades are also required. Long needle-nose pliers have long, slender jaws terminating in a fine point. They are fragile and intended for handling small parts and small diameters and for reaching into tight places. Slim-nose pliers are similar to needle-nose pliers, but the long tapered jaws are heavier at the base. These pliers are used for bending and forming small, thin wire and for holding small parts. They should not be used for bending heavy wire because the jaws may become misaligned.

Some very handy special tools for the digital technician come from a variety of places like a drugstore, hobby shop, surgical supply, and even your wife's makeup bag or sewing kit. Some of these tools are shown in Figure 10-1.

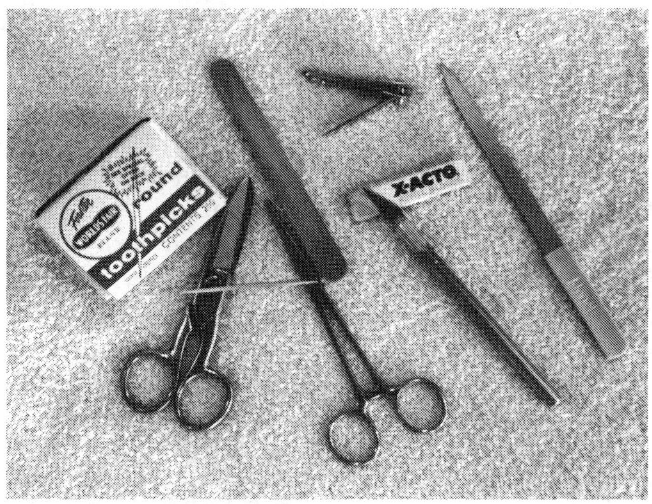


**Figure 10-1.** Special, very handy tools: fingernail clippers, hobby knife, hemostat, snips, toothpicks.

### Removing and Replacing Socketed Components on PCBs

In solid-state equipment heat and static electricity can damage components. For parts soldered in place, both heat and static discharge damage are problems. For components mounted in a socket, static discharge is the main concern. We will discuss soldering techniques and static electricity damage to components, before we actually remove and replace the component. After a discussion of electrostatic discharge damage, we discuss removing and reinstalling socketed ICs. Then, before discussing removal and replacement of soldered-in components, we discuss some soldering fundamentals.

## Zap!—Static Damage Awareness

Static electricity is a familiar natural phenomenon that, except for an occasional mild shock or the annoying "static cling" of socks in the clothes dryer, is not serious to most people. Just a few years ago, electrostatic discharge (ESD) was not a problem in electronics either. However, an increasing number of static-sensitive, microelectronic, solid-state devices just cannot take the jolt of an ESD.

The digital technician must realize the magnitude and reality of the ESD problem. Usually, a static-damaged component exhibits no visible physical damage even when examined under a microscope. Unfortunately, many electronic components can be damaged or destroyed by ESD at potentials well below a person's range of sensory perception. Both passive and active components are susceptible. The damage ranges from a slight change or degradation of a characteristic to total failure, such as a "dead" short. These total failures are easy to track down and verify. However, it is difficult to verify that wounded parts that fail later failed because of ESD damage. Thus, understanding this troublemaker is important. As a troubleshooter, you must take part in the battle against static zap.

Static electricity is electronic charge at rest on a surface; in other words, it is an excess or a deficiency of electrons on a surface. Whenever two dissimilar materials are in close proximity with one another, especially materials such as acrylic, polyester, plastic, or cellophane, one surface tends to gain an excess of electrons while the other surface loses them. The actual gain or loss of electrons is influenced by the materials themselves, and humidity and temperature. The important thing is that one surface has become negatively charged, and the other positively charged. If the materials are nonconductors, then a considerable charge can build on their surface, which could eventually dissipate through a low-resistance path or by combining with ions in the air. However, if a conductor (a conductive path) is brought near, then "zap"—an electrostatic discharge that creates heat and, with sufficient voltage, light and sound, which is what thunder and lightning are all about.

For example, when you walk across a carpeted floor, each step builds a charge that is distributed over your body. Some of the charge will dissipate; however, because there is no conductive path to fully dissipate it you could build a charge that produces a potential of 15 kV or more. Then when you reach for a metal door knob it conducts the charge back to the carpet through the door. Now, 15 kV is quite a jolt, but your threshold of feeling is such that you would probably not notice even as much as 3 kV.

Electrostatic discharges so small that you cannot feel, see, or hear them are very important to static-sensitive microelectronic devices. A metal-oxide semiconductor, for example, can be damaged by as little as 100 V. These devices are less susceptible to damage when they are installed in a circuit board, but that just means that it takes more static electricity to blast them. A 100-V zap now can later mean 100 hours of labor in lost time repairing the equipment.

By following these practices, you can nearly eliminate or greatly reduce ESD damage.

1. Use a properly equipped, static-safe workstation when handling static-sensitive components. Static-safe workstations are designed to prevent random static charges from building to a device-damaging level. They are covered by a conductive table mat that connects to ground through a 1-MΩ resistor. A conductive wrist strap connects through a swivel connector and a high-value resistor to the conductive table mat and to bare skin on your wrist. All metal equipment, including soldering irons and shelves, must connect to a common ground. The work area must be kept clear of nonconductors: common plastic, polybags, cardboard, food wrappers, styrofoam cups

and carpet are not allowed. Just lifting your feet from a carpeted floor can generate 2 kV of static electricity on your body. For this reason, a wrist strap is important.

2. Discharge personal static before handling devices. Personal static can be created and transferred by your clothing. Allow 5 to 10 s before you touch any devices or circuit boards, and then do not touch leads, pins, or traces while handling them. Clothing must never come in contact with components or assemblies. Avoid wearing nylon or other synthetic materials in a static-sensitive area. Cotton blends will produce less static electricity. Keep in mind the static charges that are generated just by separating clothes from the clothes dryer.
3. Treat all electronic parts and assemblies as static-sensitive. Though all electronic components are static-sensitive to some degree, some are much more sensitive than others. The smaller the part, the less power it can dissipate and the more likely that it can be damaged by ESD. CMOS devices are very small, so the chance of damage is great. It is wise to treat all ICs as static-sensitive.
4. Keep the replacement parts in their original static-free containers until ready for use. Do not use styrofoam packing materials around solid-state parts or circuits, especially CMOS devices. Do not slide sensitive devices over any surface. Whenever you pass ICs to someone else, be sure to touch the person before either of you touches the IC (see Figure 10-2).
5. Minimize handling as much as possible. Keep in mind that any contact or movement of a sensitive device creates an opportunity for electrostatic damage, so use an IC insertion tool whenever practical. Pick up static-sensitive devices by the body only. Leads of ICs, subassemblies, and circuit boards provide the most conductive, and therefore most inviting, paths into the device's vulnerable substrate.

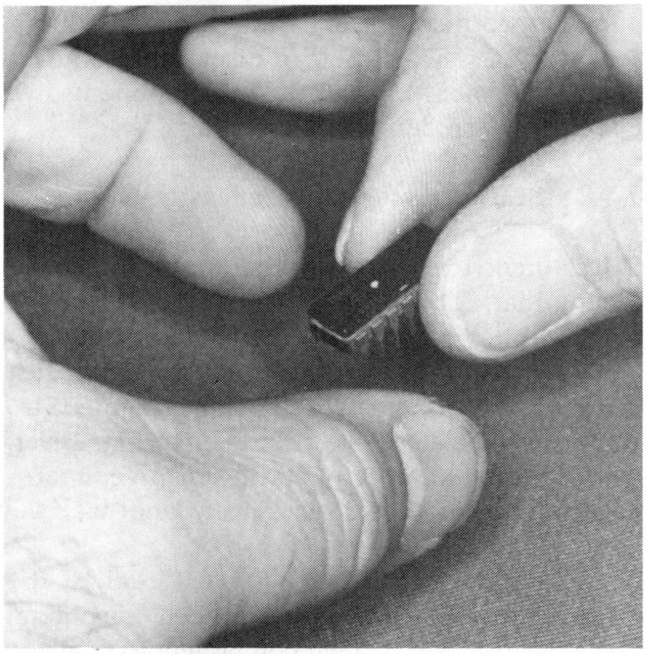

**Figure 10-2.** Passing ICs between persons.

## Removing Integrated Chips from Sockets

ICs mounted in a socket are much easier to remove and substitute than are soldered-in ICs. Thus, they are handled much more, and the chance for static damage is greater. It is quite possible that you are lifting a CMOS device when you think that you are removing a TTL chip. Because of the danger of accidental electrocution of the device, it is wise to use ESD precautions on every device. Also, it is a good practice to disconnect the equipment from the power line before components are removed or replaced. Merely turning off the power on the equipment is not enough.

To properly remove an IC from a socket on a PCB, use a DIP extractor tool. The use of an extractor is shown in Figure 10-3. The DIP extractor is a special steel tweezer with two lips that can be placed under the two ends of the IC. You can then gently pull the IC out of its socket. A small screwdriver or alignment tool can be slipped under the chip to start it out of the socket. This allows the extractor tool to get under the IC ends instead of hooking onto the socket ends.

**Figure 10-3.** Using a DIP extractor tool.

Once the chip is out of the socket, the extractor tool allows you to place it on a conductive surface, which shorts all the pins together. With the pins shorted together, a static charge cannot build up and damage the semiconductor's substrate.

The DIP extractor tool may be used to remove any size IC chip, but for a larger chip (24 pins and more) extra care is necessary. Ease it out of the socket slightly with the screwdriver, then gently rock it until it is free from the socket before lifting it up. When the chip is free, place it on its pins on a conductive surface. When using the extraction tool, connect it to a convenient ground. There is usually a place on the extractor to connect a grounding lead, such as a hole on the top. A good place to connect the grounding lead would usually be the equipment chassis ground bus.

## Replacing Integrated Chips in Sockets

It is best to use a DIP insertion tool to install an IC chip in its socket. The proper use of the insertion tool is shown in Figure 10-4. The post on the top of the tool is for connecting a grounding strip. The post is connected through the tool to the metal holder lips on the other end and also to the pin straightener in the side of the tool.

When you use the insertion tool, pull up the post to open the lips. Then with the tool over the chip, which is standing upright with its pins all contacting a conductive surface, slowly release the post so that the lips properly hold the chip. The metal lips will be contacting all the pins and grounding them. This is the safest way to handle the chip, so now look closely at the pins to see if any are bent. The pins are fragile and bend easily, and if bent and straightened many times they will break off. If any pins are bent, pick up the chip from a conductive surface with the DIP extractor tool and insert the pins into the grounded pin straightener. With the chip inserted in the pin straightener as shown in Figure 10-5, rock the chip with the extractor tool until the pins are in line and have correct spacing. It is a good idea to straighten the pins on a chip, even if they appear to be straight, before you insert it in its socket. The chip will seat much easier, and there is much less chance of bending or misaligning the pins.

After the pins are straightened, insert the chip by picking it up with the insertion tool, as described, and placing it over the socket. Be sure that pin 1 on the chip is over pin 1 of the socket. Remember, pin 1 is marked on the chip with a dot or a notch (see Figure 10-6). A good check to be sure that the chip is not inserted backwards is to notice how the other chips around the socket are mounted. Double-check at this point that the chip is not offset by one pin, as shown in Figure 10-7. It is very possible to ruin a sensitive chip by turning on the power with the pins offset on a chip.

If the chip is properly lined up and tilted to one side as in Figure 10-8, one row of pins are started in the socket. With the chip at this angle, gently press downward on the chip so that the other row of pins begin to slip into their sockets. Once all of the IC pins are started in the socket,

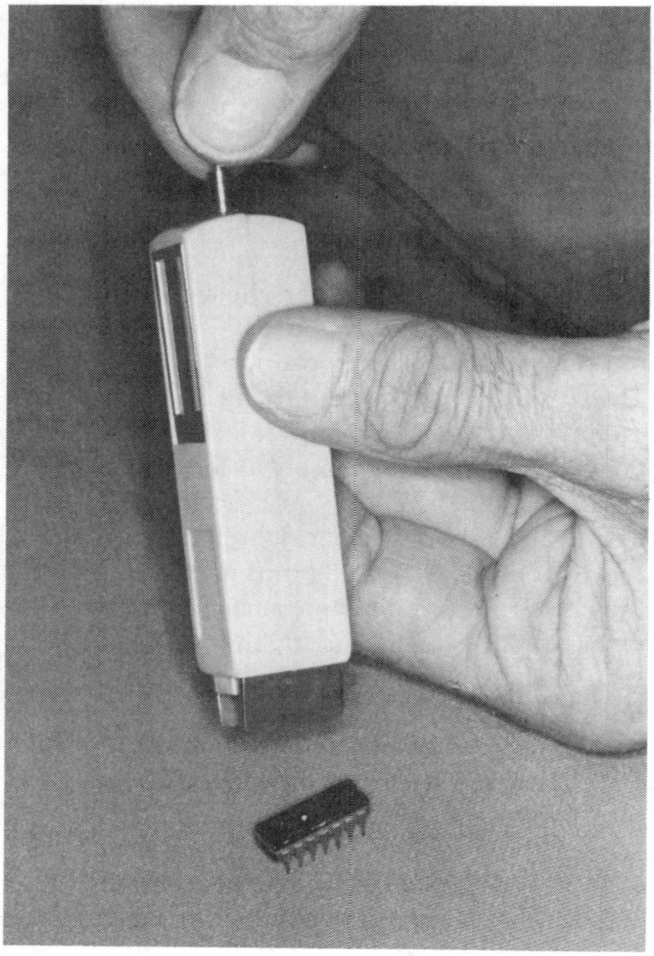

(A) Pull up the post to open the lips.

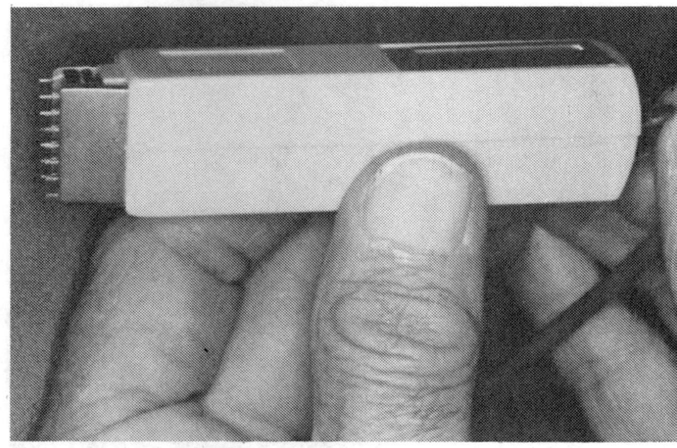

(B) Release post so that lips hold chips.

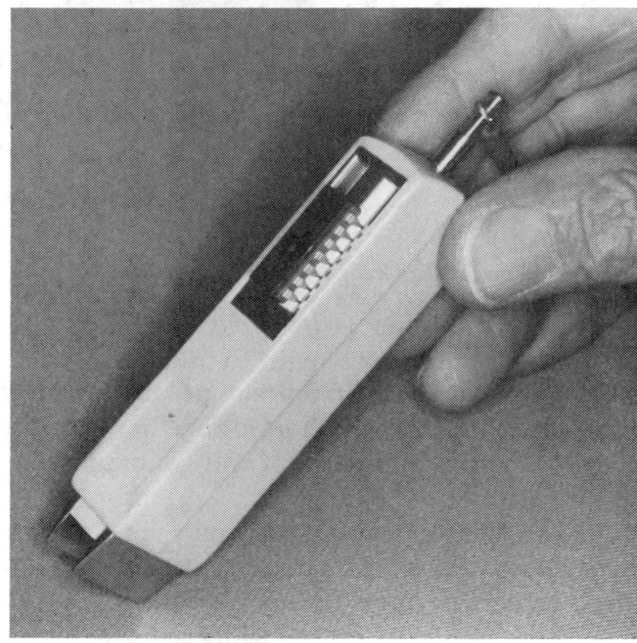

(C) Gently insert chip into its socket.

**Figure 10-4.** The steps in using a DIP insertion tool.

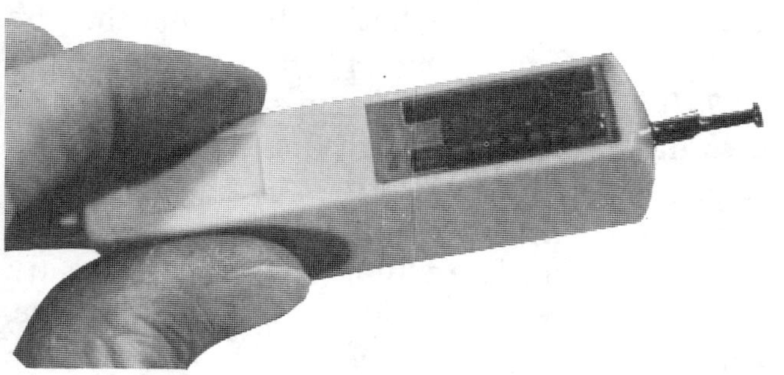

**Figure 10-5.** IC pin straightener rocked back and forth with IC extractor.

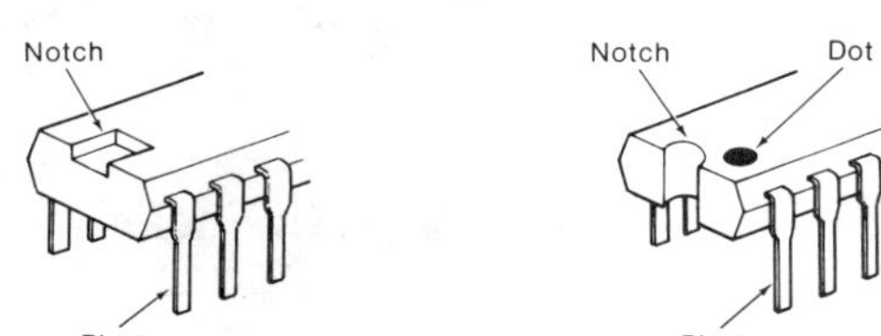


**Figure 10-6.** A dot or notch to mark pin 1 of an IC.

**Figure 10-7.** A chip mounted with the pins offset by one.

**Figure 10-8.** Inserting a chip, one row tilted.

release the insertion tool by pulling the post up on it. Press the chip firmly down in the socket until it seats all the way. Be careful not to flex the board too much. It may be necessary to support the board under the socket as you press the chip in place.

## Removing and Replacing Soldered-In Components

Removing and reinstalling ICs, transistors, diodes, and other components that have been soldered into a PCB requires both knowledge of the soldering process and some special skills and techniques.

To solder sensitive ICs to the delicate pads of a PCB, it is vital to choose an iron of the proper size and wattage and to develop the skills necessary for a quality finished job. Soldering irons vary in size, shape, and wattage. Too large an iron contains an overabundance of heat in its large storage space. This excess heat travels quickly into surrounding areas, creating the danger of overheating components or adjoining connections.

Manufacturers normally rate their irons in watts, 18–65 W being typical for electronic applications. It is difficult to compare exact tip temperature with wattage rating because the length and diameter of the tip largely influence the tip temperature. An 18–25-W iron is generally used for heat-sensitive and small electronic components. A 40–65-W iron is used for large terminals and fasteners.

When soldering components to PCBs, apply a small amount of solder to the iron tip for more efficient heat transfer. Place the cleaned and tinned iron tip on one side of the connection point and fresh rosin core solder on the opposite side of the joint members, away from the hot iron tip. This procedure ensures that the iron tip will heat the connection point, which, in turn, will heat and melt the rosin core solder. Remember, for good wetting action, the metal to be soldered must be as hot as molten solder. It is the connection point that actually melts the solder. The solder will heat relatively slowly and melt and flow around the joint.

### Component Removal

Once the suspected defective component is located, it must be unsoldered and removed from the board. This involves heating the joint, one of the most critical operations in repair. Too much heat can quickly destroy a good component or cause the circuitry to become unbonded from the board. So, most desoldering operations have to be done rapidly, usually in no more than 2–3 s. Some factors that affect how rapidly a joint heats up are

1. The relative thermal mass of the joint: the joint is affected by the size of the component leads, whether the board is single- or double-sided, plated-through or not plated-through, and other things.
2. Surface condition: the joint may have an oxide or a conformed protective coating that creates a barrier to the heat flow.
3. Thermal linkage: the area of contact between the heated tip and the joint greatly affects the transfer of heat, and the proper choice of iron tip is important.
4. Amount of solder: the solder itself is the principal transporter of heat to the joint. If there is not enough solder, then the iron will take too long to transfer heat and damage to the board is likely. Adding a small amount of solder as the iron tip is applied will speed up the heating process.

Soldered-in components can be properly removed from boards by using one of several appropriate methods. A method that is *not recommended* is to heat the solder until molten, and then strike the board on the benchtop to shake the solder loose. The shock can easily damage the components, and the hot splattered solder can burn the board and create shorts.

An acceptable and widely used method is *wicking*. All you need is a piece of wire-braided copper material that has been saturated with flux. Place this braid on the top of the solder joint and then bring a hot, tinned soldering iron down on top of it. The combination of heat, molten solder, and the air spaces in the wick creates a capillary that draws up the solder. You may have to move the braid a time or two for complete absorption of the solder from the joint. Proper use of the braided wick for solder removal is shown in Figure 10-9. It is important to slide the iron and the wick off of the joint at the same time. Be careful that the joint is not overheated, which will damage the circuit board. Cut off used portions of the braid and use a new end of fluxed braid each time it is applied. Don't push down hard on the iron because too much pressure could lift the pad. This method is good for surface solder joints, but not for plated-through holes. It can leave solder in the hole, and reheating may lift the pad.

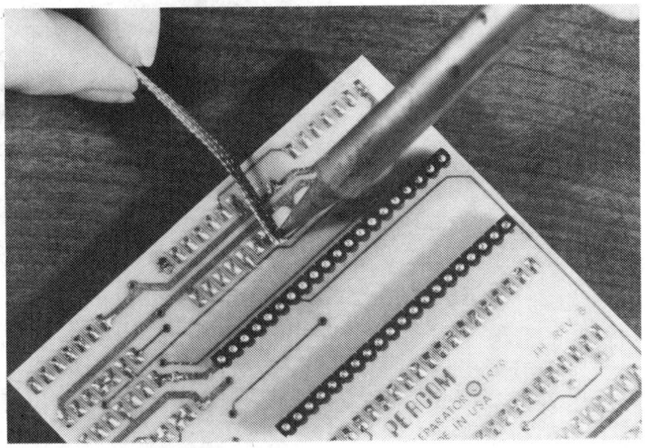

**Figure 10-9.** Braided solder wick in use.

The *vacuum pulse* or *vacuum pull* method of solder removal is the most widely used. The vacuum is created by either a rubber squeeze bulb, like a syringe, or a spring-loaded piston pump called a *solder sucker*. This provides positive vacuum forces to overcome the capillary action of the plated-through hole. A solder sucker in use is shown in Figure 10-10.

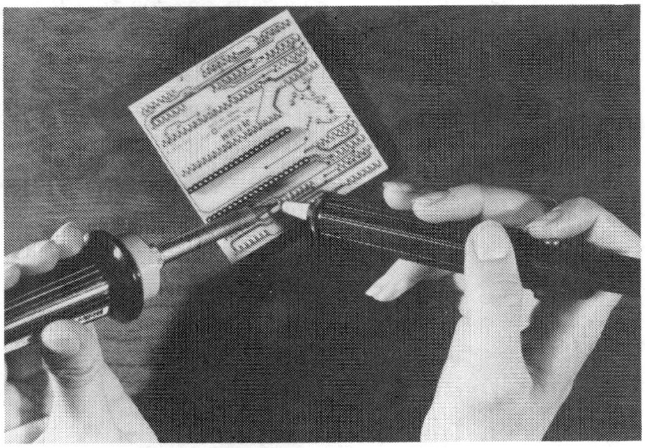

**Figure 10-10.** Solder sucker in use.

To use the squeeze bulb, heat the joint with a soldering iron. Opposite the iron, squeeze the bulb with the tip as close to the joint as possible without touching it. When the solder is at molten temperature, release the bulb quickly and let the bulb expand to normal size. The molten solder will be drawn into the solder reservoir. When this is done, remove the iron and syringe. After you have desoldered several joints, remove the syringe from the bulb and clean out the old solder.

The solder sucker is used in a similar way. The trigger activates a spring-loaded plunger that causes the vacuum to pull the molten solder into the orifice. Reloading the spring plunger expels the solder that was drawn into the orifice during the vacuum stroke. Unless the solder sucker is a low-static type, it is not recommended for removing static-sensitive parts.

On the syringe and plunger types, you can prolong the tip life by not pressing the tip orifice into the melted solder. Hold it only close enough to the solder for proper removal.

The best method for removing multileg parts is the *solder vacuum* method shown in Figure 10-11. This consists of a hollow-tip soldering iron connected to a vacuum source. The vacuum source is controlled by a foot pedal. You need not reload after each operation, a real convenience if you must desolder many joints. To use this semiautomatic desoldering tool, place the hollow tip over the component lead. As the solder melts, move the tip in a circular pattern to keep the component from bonding to the side of the hole during the extraction process. When the solder becomes completely fluid, start the vacuum, which extracts the solder and cools the connection at the same time. This process occurs so quickly and heats such a small area that there is usually no heat damage. If available and economically justified, the solder vacuum method is best for removing any part.

The *heat-and-pull* method shown in Figure 10-12 should be used only on two-lead parts such as diodes, resistors, and capacitors. To remove a component, grip one of the component leads with long-nose pliers. Touch a soldering iron to the lead at the solder connection. Do not lay the iron directly on the board. When the solder begins to melt, pull the lead out quickly but gently. Repeat this action on the other lead. Then heat the pad

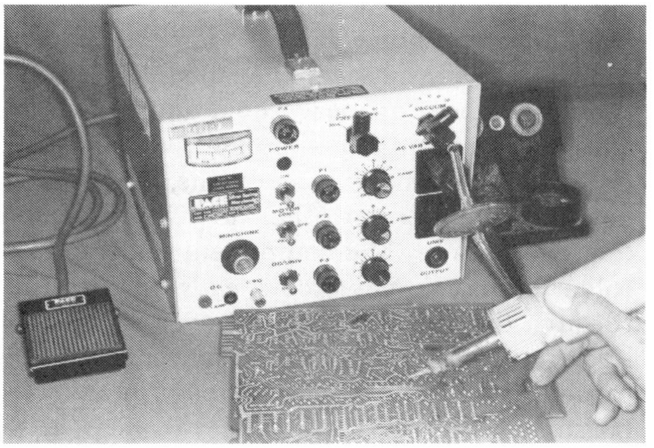

**Figure 10-11.** Vacuum pump semiautomatic solder remover.

where the part was removed and remove solder from the hole. Bend the leads of the new component to fit the holes in the board. If you replace the component while the board is mounted in the instrument, cut the leads so that

**Figure 10-12.** Heat-and-pull method of component removal.

they just protrude through the board. Insert the leads into the holes in the board so that the component is firmly seated against the board. If it does not heat properly, heat and add solder and gently press the component into place. To protect heat-sensitive components, use a pair of needle-nose pliers or other heat sink and hold the lead between the component body and the solder joint. Clip any excess leads that protrude through the board. Clean the area around the solder connection with a flux-remover solvent. Be careful not to remove information printed on the board.

The *cut-part-lead* method of component removal shown in Figure 10-13 consists of cutting the lead between the body of the part and the solder connection. When the connection is heated, use needle-nose pliers to pull the lead out of the hole. Remove solder from the hole and add the new part. This method of removal is very good for removing machine-inserted parts, such as resistors, because of crimped leads on the back of boards. Resistors are inexpensive—the cost of a new resistor being much less than the cost of a circuit board damaged by heat.

**Figure 10-13.** The cut-part-lead method of component removal.

The easiest method of extracting ICs is the *clip-out* method. Use this method only if you are very sure you have identified the bad IC, and you have a good supply of replacement IC parts. In the clip-out method, cut off each pin as close as possible to the body of the IC. Remove the body, leaving all of the pins of the IC still soldered in the board. Remove the pins one at a time by applying heat with a soldering iron and pulling the pins out with needle-nose pliers. Other methods may be faster, but the clip-out method is easy to learn and there is little chance of overheating and damaging the PCB. Once the IC has been clipped out and the pins removed, clear each hole on the board of solder. A simple way to do this is to push a round toothpick through the hole. A more efficient method is to use a vacuum device, such as a squeeze bulb or solder sucker.

As mentioned in Chapter 2, you may have to lift a part lead or a transistor lead to isolate electrical problems. However, it may be safer to just cut the part out rather than to risk ruining a circuit board pad or run. If you are sure that a component is bad and you do not care what happens to it, clip its leads and removal will be much easier.

### Component Replacement

After you have removed the suspected bad components, prove that it really was bad by substituting a known good component in its place. If the new one eliminates the trouble, then it is fairly conclusive that the old one was defective. Sometimes it is a good idea to temporarily install a component in a circuit board without soldering it in place.

*Temporary Replacement of a Part*—A troubleshooting technique in vacuum tube days was to replace the suspect with a tube known to be good. But tubes were not soldered in the circuit—they were always socketed. It is a good idea never to unsolder and resolder the same chip twice. However, because of an occasional design error, a particular chip may repeatedly fail. In that case, it is good practice to install a socket whenever you take out an unsocketed chip. Then you will never again have to unsolder and resolder that particular chip. It is much easier to do this on PCB runs and pads, and the IC may be temporarily substituted when troubleshooting.

Instead of installing a socket in the board, you can temporarily install an IC in the circuit if it has plated-through holes by "toothpicking." This method is used by some effective technicians and works well. After removing the suspected IC and cleaning away the solder from the holes, insert the new IC in the holes, being very careful that it is in the right direction. The new IC will probably be ruined if it is inserted backwards. Hold the chip firmly against the board and insert toothpicks into the holes beside the IC pins, forcing them in solid contact with the plated sides of the hole walls as shown in Figure 10-14. The holes must be plated-through for making good contact with the pins, and the technique should not be used for single-sided boards.

**Figure 10-14.** "Toothpicking" an IC to temporarily substitute a new one without soldering it into the circuit.

As we discussed in Chapter 2, a good way to temporarily replace an IC suspected of having an internally *open* pin is to piggyback a new IC on top of the suspected IC. If piggybacking cures the trouble, replace the bad IC. Be sure that you are piggybacking the same type of IC, and be careful to watch the pin orientation. You may also need to squeeze the pins of the new IC together to ensure a good electrical contact. Figure 10-15 shows the piggybacking technique.

**Figure 10-15.** "Piggybacking" an IC that is suspected to have an open.

*Permanent Replacement of a Part*—Remember that soldering components onto circuit boards requires considerable caution due to the heat sensitivity of both boards and components. Skilled technicians have learned to use a very light touch when permanently soldering in the replacement part. Never push the iron tip down on the pad, merely rest on it. Board damage comes mainly from heat combined with pressure. Most replacement parts will be from one of three groups: axial-lead components (resistors, diodes, capacitors), TO case component group (transistors, ICs), and dual in-line pins (DIP ICs) group. We will consider each in turn.

Clean the leads of resistors, diodes, capacitors, and other axial-lead components before soldering. Use a very light, noncontaminating abrasive, even a soft rubber eraser, on the component leads and the pads. Clean gold-plated leads only with solvent and tissue to prevent removal of the plating by an abrasive.

When bending the leads of the components, use a tool that will not damage them, such as pliers with nonserrated jaws. If pliers with serrated jaws are used, cover the jaws with heat-shrink tubing or similar material. After bending, and before inserting the component, wipe the leads again with solvent to remove any oil or grease left from your fingers.

For straight-through (or clinched) lead termination, cut the lead to a length equal to the pad diameter and solder the joint. Remember to bring the iron in contact with the lead and the pad and apply the solder to the connection. Then remove the iron, sweeping it over the end of the lead to ensure that the solder covers it.

The TO package is used for transistors and ICs. Transistors with three or four leads are replaced in the same way as the axial-lead components. In ICs with many leads (10 or more), it is difficult to line up the leads and holes. A trick that helps tremendously, and even makes three-lead components mount easier, is to hold the case of the device and rotate the leads, cutting each one so that it is slightly shorter than the lead next to it. Then put the longest lead in first and rotate the case to align the next lead in turn. When all the leads are in place, cut them to proper length and solder.

The DIP, by far the most popular, has two rows of formed flat leads. The leads of the DIP come slightly sprung outward, so you must form the leads so that they can be easily inserted into the holes. A pin straightener is the best way to do it. Then pick up the chip with the IC insertion tool and be certain of the proper orientation. Place the chip into the holes on the board. Clinch two pins on opposite corners outward just enough to hold the chip in place for soldering.

The soldering iron should be about 18 W with its tip well matched for the job. Apply the tip to one side of the joint, and the solder to the other side. In soldering a DIP in the circuit board, first solder opposite corners and then solder alternate pins from one side to the other in order to minimize heat buildup in any one area. With single-sided boards, the procedure is similar to that for soldering axial-lead components, with the iron in contact no more than 2 s. On double-sided boards, the maximum time is 3 s. The joints should appear properly wetted and have solder fillets on both sides of the board. There should be no air pockets, flux pockets, or voids where contaminants could accumulate.

## PCB Damage Repair

Now we consider some surface repair problems that may be encountered on single-sided and double-sided boards. Keep in mind our goal—nondestructive repair that seeks to equal or even surpass the quality and reliability of the original product.

Two preliminary and basic considerations in each type of board repair are (1) avoid excessive heating, and (2) always clean and pretin the work well before starting. Remove all contamination and loose foreign matter at the onset, and wipe the surfaces clean with solvent.

To repair a damaged PCB when a portion of the run is broke and/or lifted up from the base material, a wire jumper is used to reestablish the run. The run may be surface mounted or extended through the board's base material. First cut away all unbonded circuitry; then remove any burned material with an abrasive tool. Brush away the residue and wipe the area with solvent. Next, clean the run on each side of the break about 1/4 an inch from the ends, and pretin. Since the replacement conductor will reflow solder into place, it must have its ends pretinned and cleaned. Then put liquid flux on the joint area. If you use round bus wire, flatten it slightly with pliers to prevent it from rolling around on the board. Now cut to a length that is at least 1/4 inch longer than the break. Position the wire along the break with tweezers and solder in place. After the joint has cooled, remove the flux with a solvent. Carefully inspect the repair for quality.

Wire jumpers of the right size and properly used work well in most cases. But a better method is to replace a portion of the conductor with flat material. The material should be as close as possible to the original in dimension and appearance. You can obtain it by cutting a piece of copper sheet, removing a similar size conductor from a scrap board, or using one of the commercially available frames of pretinned track material. These frames of PCB trace, pad, and other patterns are available at most electronics distributors.

## Intermittent Problems

An intermittent problem is one that depends on variable conditions in a circuit or piece of equipment. It is not present at all times; that is, it occurs at regular or irregular intervals. The intermittent problem can be exasperating. The unpredictability of an intermittent problem makes using logical troubleshooting techniques nearly impossible. The key to troubleshooting this elusive type of defect is to make the symptom stay with you long enough to pinpoint its cause.

The intermittent trouble can usually be classified into one of four classes: thermal, electrical, mechanical, or external (originating from an external source).

### Thermal Intermittent Problems

When an intermittent is thermally induced, the trouble shows up after the equipment has been operating for some time. Then, when the chassis is out of its cabinet on the workbench, and therefore in a cooler environment, the equipment behaves normally again. To isolate the trouble, you must create a heat environment. You can cover the chassis with a large carton or blanket and put a 60-W lamp inside. Be careful not to overheat the chassis and thus damage good components.

When the intermittent has been localized to a board or section, apply heat more directly to isolate the problem. A commonly used device for this purpose is a heat gun or a hair dryer with a nozzle for directing heat to a specific component. Here, alternately heating and cooling the components in an isolated area will usually prove very effective. Most components can be cooled rapidly with an aerosol coolant spray. If there is a thermal intermittent inside an IC or other component, cooling it with cold spray will probably cause it to work properly again. Heating the suspected component may induce the problem again. Repeat this process until the defective component is isolated.

Exercise discretion in using this technique, to avoid severe thermal stress that can damage good components. A 2- or 3-s "blast" of heat or cold should be sufficient to locate the thermal intermittent. If the intermittent does not respond to these abrupt temperature changes, do not cook and freeze it to death. In some cases, the component will permanently open or short, making it easy to locate. Otherwise, try a different approach.

### Electrical Intermittent Problems

When a trouble seems to be thermal but does not respond to the preceding approach, it may be sensitive to voltage or current variations; that is, it could be a condition affected by line voltage or power supply fluctuations, surges, or transients. The object, once again, is to make the equipment misbehave by whatever means possible before attempting to find the cause.

An effective approach is to increase all voltages and current throughout the equipment by raising the line voltage with a variac. Do not raise the voltage above 128 V. Avoid prolonged increased voltage and watch carefully for any signs of overheating.

In some cases, the intermittent can be induced by reducing the line voltage with the variac. With some equipment, the voltage regulator circuit makes it difficult to raise and lower voltages in this manner. This method may also point to the voltage regulator as the origin of the intermittent.

### Mechanical Intermittent Problems

When the equipment is susceptible to movement or vibration, attempt to recreate the problem by tapping it, especially the PCB, at various points. Use a solid object, such as a screwdriver handle. Try to localize the intermittent by moving the tapping point around the equipment and reducing the intensity of the tapping. When the problem is localized, use a lighter tool for isolating it, such as the eraser end of a pencil or a plastic alignment tool. A gentle push on the PCB may be useful for finding a hairline crack in a trace or a bad solder connection. A large magnifying glass with its own light is sometimes quite useful here.

A bridge or short from a splash of solder is a common cause of an intermittent problem, as are bare wires that may be shorting together. Separate any wires that look at all suspicious. A component could have an intermittent where its lead connects just inside its body. Try using pressure with a probe to put stress on its lead. Touch up questionnable connections, especially where large components connect to the PCB in the suspected trouble area. If necessary, as a last resort, rework each solder joint.

## Externally Induced Intermittents: Noise and Interference

Much of the dramatic change and progress in digital electronics has come from microminiaturization. One of today's microprocessor chips has the equivalent of nearly 100,000 transistors. Along with this large-scale integration has come the problem of sensitivity to overload voltage, especially to MOSFET devices. A voltage spike of less than 1 V can, in some instances, ruin a MOS device. These overloads can come from static electricity, the electrical power line, or electromagnetic radiation.

### The Electrical Power Line

Let's look at the typical workbench situation. The PCB under repair is in direct contact with a soldering iron or a desoldering tool. It is connected to a power line outlet, which in turn is connected to the main source of power. There is a direct physical connection all the way through the system, and voltage spikes generated in equipment elsewhere in the power system can pass directly into the board and into the sensitive components. Switching highly inductive loads, such as motors and transformers, on and off can be a major cause of these spikes. Using a zero-power switching method is the best way to eliminate this problem.

Another source of the problem is the power line itself. Brush motors, relays, and arc or spot welders are notorious sources of these spikes on the power line. Using a proper filter between the noise source and yourself is a good way to stop this interference and damage.

Positive grounding is one of the key methods of preventing damage to sensitive components by this type interference. Make sure that your power system is a properly wired three-wire system with constant polarity. Also, be sure that the soldering equipment selected has a grounded tip and is plugged into a power outlet with proper grounding and polarity. In some situations, you may also want to positively ground the board you are working on. However, a second ground may cause a ground loop with the board and soldering iron at different potentials.

### Electromagnetic Radiation

The other form of energy that can cause damage is electromagnetic radiation. It is a field of energy that makes radio and television possible. For radio and television to work, both the transmitter and receiver require antennas. The closer the receiver antenna is to the transmitter, the more energy that will be induced in it. This type of field is ground referenced. Thus, if the receiving antenna is grounded, no energy will be induced into it. Transformers act somewhat like transmitters in that they emit radiation of a fluctuating magnetic field. The soldering iron can act like a receiving antenna—that is, it can pick up the induced energy and carry it into the PCB being repaired. To eliminate this electromagnetic-induced energy, ground the soldering iron.

# Safety Practices

The electronics technician works with many complex devices and electrical equipment that can be hazardous to both the technician and the equipment unless certain safety practices are followed.

## Safety Practices for the Technician

Accident reports show that careless use of electricity can lead to burns and shock, and more serious shock-related injuries such as cuts or falls. This discussion is intended to help you avoid such accidents.

The small voltages and currents that are worked with in everyday situations cause more accidents than the high voltages and currents of the power company's transmission lines. The real measure of a shock's intensity is determined by the amount of current in milliamperes forced through the person's body. A current of 10 mA or more is capable of producing painful and severe shock. The amount of current through your body is determined, according to Ohm's law, by the amount of your body's resistance and the voltage applied. The resistance varies according to the points of contact, the length of exposure to voltage, and whether your skin is wet or dry: the dryer the skin, the higher its resistance.

Currents between 100 and 200 mA can be more dangerous than larger currents because they cause the victim's heart to start an uncoordinated twitching called *fibrillation*. When this twitching occurs, a rescue unit's defibrillator is the only way to save the victim. In some circumstances, a current of only 10 mA can cause this twitching. So, because of these variables—point of contact, length of exposure, moisture of skin—low voltages can be just as harmful as high voltages. Understanding these facts should make it easier for you to recognize the importance of following some basic safety practices.

One of the most common causes of electrical shock is a "hot" chassis or instrument case caused by a defective or defeated ground on the power cord. Before plugging in the equipment's power cord and before doing any troubleshooting, perform a continuity test. Make sure that the green ground lead of the power cord is properly grounded. It is there for a reason, so don't disconnect or defeat it. Measure the resistance from the ground pin on the power cord plug to the equipment chassis. It should measure 0 Ω.

Since current from one hand to the other hand goes through the heart, keep one hand behind you or in your pocket when checking high voltages or line voltages. Keep your free hand away from the instrument. Accident reports prove that this is a very good practice to follow.

Do not poke around in equipment with conductive objects, such as a pencil or screwdriver. Use a plastic or fiber instrument when probing around in a live circuit. More than once, a metal flashlight has come in contact with voltage and caused an accident.

Keep your work area clean and uncluttered. Clean up any spilled liquids immediately.

Working around a CRT brings a special set of safety hazards. Even when the instrument is turned off and unplugged, there is potentially high voltage stored on the CRT and in its power supply. If you must disconnect the anode lead during troubleshooting, then connect both the CRT anode button and the anode lead to ground for several minutes to completely discharge them.

Capacitors can store a charge for hours after the power cord has been disconnected. The capacitors in switching power supplies present a particular hazard. They all must be discharged at once. A resistor with clip leads connected across each of the capacitors will discharge the capacitor in seconds. The proper value will depend on the voltage rating of the capacitor. An encased (insulated) 1000-Ω, 2-W resistor is a good general choice.

Another area of potential hazard is the line fuse holders. Even with the power switch turned off, there is power line voltage available at the fuse terminals. It is good practice to remove the line plug from the power outlet, connect the appropriate test leads, reconnect the power plug to the power outlet, and make the measurement. Remove the plug again to disconnect the test leads.

By following these safety practices and using common sense, you can avoid shock hazards and safely troubleshoot equipment.

### Safety Practices for the Equipment

Any repair work would be mechanically sound and electrically reliable. It should not have a degrading effect on overall appearance. During troubleshooting and repair, some components and equipment require special handling.

During soldering, many heat-sensitive components should be protected against heat damage, including solid-state devices as well as hermetically sealed components, which should be heat sinked between the solder joint and the body of the part.

The leads of the component should not be subjected to stress or strain. This can weaken internal bonding and break hermetic seals, paving the way for future intermittent troubles. Hold the component with nonserrated-jaw pliers between the body and the place on the lead where the bend is to be made.

Most PCB damage occurs during part removal. It may be better to clip the part out and remove the leads to reduce the chance of damage to the board. Use extra care when soldering wires. Melted or burned insulation, nicked wires, and loose strands are unacceptable. All flux is corrosive and must be removed with an approved solvent. Do not use freon solvent.

Keep your work area clean and uncluttered. Keep tools in their proper places. It is all too easy to pick up a screwdriver instead of the proper insulated fiber alignment tool if things are not in place and easily found. Using the wrong tool is a common way of inducing extra trouble into the equipment. Bits of paper can be drawn into filters by fans in equipment and completely block the airflow, so keep paper away from the blower grills and keep all unnecessary papers off of the workbench. Spilled liquids are also potential dangers to electronics equipment and can make a terrible mess on a PCB, keyboard, or disk drive. Coffee cups, drink cans, and glasses should be kept away from the equipment. This bit of advice can also be considered as preventive maintenance.

## Operational Checkout

Competent troubleshooters always check to make sure the reported trouble is cleared. But they also check to see that the total system is functioning normally after repair. It is all too easy to insert a new trouble that is totally unrelated to the originally reported trouble. This can happen at any time during the troubleshooting process—during disassembly, isolating and testing, while eliminating the trouble, or during reassembly. For example, you may break a wire in a cable assembly or bend a pin on a plug or an IC chip while reinserting either of them in its socket. Therefore, a final checkout of the total system after repair and reassembly is a necessary part of the total troubleshooting sequence.

## Documentation

A well-documented log of the equipment's maintenance history can be of great benefit to the troubleshooter. Quite an impression was made on this writer some years ago by an excellent technician who owned a television shop. He kept a detailed record of every repair on each Sams Photofact® schematic folder. Many times, this record showed him trouble patterns with a particular make and model of television. He did not have to rely on his memory alone to recall uncommon problems. He was the "expert," as described in Chapter 2, called on by other television technicians for help with their "dogs."

Completing the paperwork, filling out the forms, and logging the maintenance may seem like an unnecessary nuisance to the troubleshooter. It may be a nuisance, but it is not unnecessary. Well-kept records of maintenance of equipment can be very useful to those who will troubleshoot the equipment in the future. And often the documentation goes well beyond that. It can be a part of management's data for planning or purchasing future equipment, and it is vital for spare parts inventory control. Good troubleshooters, therefore, understand the need to "fill out the paperwork."

A final note on this subject of documentation: the technician should be correct and consistent in the use of technical terms in order to aid in communication with those who will read the report. Avoid jargon and be professional.

# Preventive Maintenance

The rapid increase in digital-based equipment in a short period of time has caused us to forget the importance of preventive maintenance. Only a few years ago, digital electronic equipment meant enormous floor computers that thrived in regulated environments. The temperature, humidity, and line voltage were stringently controlled. The air was clean, and the operators wore white, static-free jackets. Although mainframe computers today might get the same respect, almost every personal microcomputer, industrial controller, and digital equipment of all varieties are subjected to a savage jungle of environmental beasts. The mainframes never encountered cigarette smoke, coffee and soft drink spills, or cupcake and doughnut crumbs. Neither were they subjected to vibration, extreme humidity, and temperature variations. Digital electronics users today would do well to adopt the preventive maintenance attitude of an earlier day.

Preventive maintenance (PM) is the process of clearing troubles before they happen—a process that good troubleshooters perform as regularly and thoroughly as time and policy permit. It saves time and money in the long run by preventing downtime, decreasing the frequency of malfunction, and increasing the life of the equipment by preventing the extensive damage that a failure can sometimes produce.

The skills needed to perform effective PM will vary greatly, depending on the equipment. However, highest on the list of skills for any technician doing PM is care—the ability to do PM without damaging the equipment or inducing a trouble into it. The mechanical type of equipment, such as floppy disk drives, stepping motors, and their associated gear boxes and printers, may need lubrication and adjustments without dripping oil or smudging up the inside or outside of the equipment case. The adjustments must be made without scratching, scarring, or scuffing the equipment. These are the things that tear down in minutes any goodwill built in months of using PM. It turns operators and managers alike against the entire philosophy of PM.

Many have developed the idea that PM is a thing of the past. They have two good reasons to think so:

1. Equipment today, especially nonmechanical or purely electronic equipment, tends to wear well. When it fails, it is often the result of human error or adverse operating conditions.

2. Performing PM inside modern digital equipment passes hazards to it: wires are pulled, boards and edge connectors may be flexed and bent, which may produce a microscopic crack, and probing with test leads can cause failures in the process of trying to prevent them.

Therefore, serious thought should be given to the option of doing PM or doing nothing with digital equipment that is working properly. However, certain areas in digital equipment maintenance lend themselves well to PM. As mentioned, almost any mechanical device responds well to certain basic things. First on the list is simply a matter of keeping it clean. Dust and dirt are the main culprits.

For a printer, PM is a matter of keeping it clean, inside and out. It can be thoroughly cleaned by removing the cabinet and carefully vacuuming out the inside with a small vacuum cleaner. This can also be accomplished with a pinpoint (antistatic) compressed gas duster, which will dislodge dust and lint from hard-to-reach surfaces. This duster can be used in many PM applications. Watch out for small levers, cams, wheels, gears, and cables. The printhead and other moving parts may be hot from use and should be allowed to cool before cleaning. The printhead can be cleaned with denatured alcohol on a small foam swab available for that purpose. Do not use cotton swabs: they leave a trail of lint behind. The carriage rails can be cleaned by the same method, but should be relubricated according to the manufacturer's instructions. CRT screens need to be cleaned with an antistatic solution. Antistatic solutions generally are helpful for most components and the surfaces around them.

Only the proper lubricant should be used on the mechanism of any electronics equipment. Improper lubricants can cause a gummy or sticky contaminant buildup that will void any good that the PM was to accomplish. Rely on a quality and comprehensive set of cleaning and lubricating products that are designed for specific tasks.

Use, and encourage the use of, consistently high-quality products with your electronics equipment, especially printer ribbons, diskettes, and tapes. Using bargain supplies could cause poor print quality, jams, premature wear, and loss of head alignment and data in disk drives.

By establishing and encouraging an effective PM procedure, you make a considerable positive contribution to both your own success and that of your company and customer.

# Index

## T



## V



## W



## Z

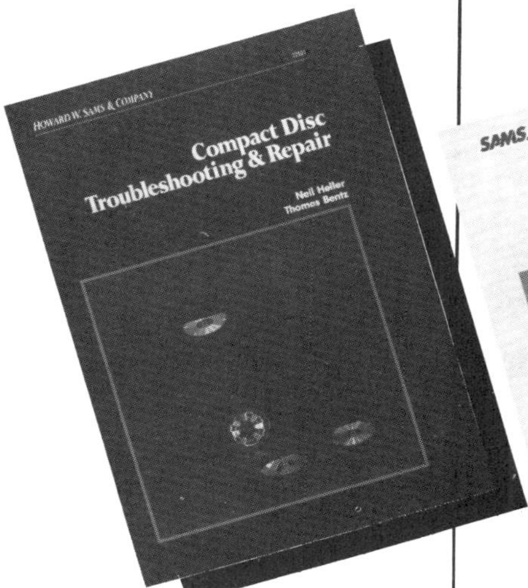


## Compact Disc Troubleshooting & Repair

*Neil R. Heller*

This book discusses much more than how to repair a compact disc, it is a concise and complete introduction to the field of digital signal processing and compact disc players. The book describes in detail the hardware and technology involved in this expanding new field.

Topics covered include:

- The Audio Signal and Signal Processing
- Basic Digital Theory
- The CD Modulation System
- Construction of the Compact Disc
- The Laser Pickup
- Detailed CD Circuit Descriptions
- Adjustments and Troubleshooting
- The CD-ROM
- Glossary of Terms

160 Pages, 8½ x 11, Softbound
ISBN: 0-672-22521-2
**No. 22521, $19.95**

## Commodore 64® Troubleshooting & Repair Guide

*Robert C. Brenner*

Repair your Commodore 64 yourself, simply and inexpensively. Troubleshooting flowcharts let you diagnose and remedy the probable cause of failure. A chapter on advanced troubleshooting shows the more adventuresome how to perform complex repairs. Some knowledge of electronics is required.

Topics covered include:

- Theory of Operation and an Overview of System Structure
- Preventive Maintenance Tips
- Basic and Advanced Fix-It-Yourself Procedures
- Repair with Few or No Tools
- Complete Appendices of All Necessary Technical Data

256 Pages, 8½ x 11, Softbound
ISBN: 0-672-22363-5
**No. 22363, $19.95**

## Commodore® 1541 Troubleshooting & Repair Guide

*Mike Peltier*

This guide provides troubleshooting and repair procedures for the Commodore 64® or VIC 20™ with a 1540, 1541, or 1542 disk drive. It is comprehensive and detailed, offering easy-to-follow, step-by-step instructions. It presents the theory and general operation of the disk drive and points out some common disk drive problems. The diagrams will help readers isolate their disk drive problems.

Easy-to-understand circuit diagrams and simple instructions written in a clear, concise style allow you to make most repairs with few or no tools.

Topics covered include:

- Description of the VIC-1541
- Theory of Operation
- Alignment
- Preventive Maintenance
- Basic Troubleshooting and Repair
- Advanced Troubleshooting and Repair
- Advanced Theory of Operation
- Appendices: Technical Data, Fabrication of Video Detector, MOS Handling Precautions, Standard IC Pin Numbers, Description of Disk Controller IC

200 Pages, 8½ x 11, Softbound
ISBN: 0-672-22470-4
**No. 22470, $19.95**

## How to Maintain and Service Your Small Computer

*John Stephenson and Bob Cahill*

Simple repairs can be diagnosed without a trip to the repair store. This book shows you some easy maintenance and operating procedures that will likely reduce problems and down-time. If the problem is more major, then you can apply the skill learned here when dealing with the repair shop for professional servicing. Some basic electronic knowledge is required.

Topics covered include:

- Today's Equipment—A Service Profile
- Servicing Your Equipment—the Options
- Environment and Operating Procedures—Reducing Potential Problems
- Preparation—Skills, Documentation, and Tools
- Troubleshooting Strategy and Skills
- Troubleshooting
- Disk Drives
- Servicing Printers
- Troubleshooting the Main Computer and Keyboard
- Modems, Monitors, Cassette Recorders, and Game Controllers
- Troubleshooting Power Supplies
- Appendices: Resistor Color Code, IC Manufacturers' Numbering Systems, Mail-Order Parts Dealers, Pinouts and Base Diagrams, and Troubleshooting Chart

224 Pages, 8½ x 11, Softbound
ISBN: 0-672-22016-4
**No. 22016, $17.95**

**Visit your local book retailer, use the order form provided, or call 800-428-SAMS.**

## John D. Lenk's Troubleshooting & Repair of Microprocessor-Based Equipment

*John D. Lenk*

This text provides a simplified, practical system of troubleshooting for the many types of microprocessor-based electronic devices. Concentrating on the basic approach to troubleshooting, the book includes numerous procedures and "tricks" that can be effective in diagnosing, isolating, and locating faults in microprocessor circuits.

Topics covered include:

- Microprocessor Test Equipment
- Problems in Troubleshooting Microprocessors
- Troubleshooting the Gate and IC
- Stimulus-Response Testing
- Vcc and Ground Shorts from Unexpected Causes
- Flip-Flop Troubleshooting
- Register Troubleshooting
- Current Flow Analysis
- Wired-AND and Wired-OR Troubleshooting
- Using the Logic Analyzer
- Step-by-step Procedure for Troubleshooting the VCR

John D. Lenk is one of the industry's foremost authorities on electronics and troubleshooting.

176 Pages, 8½ x 11, Softbound
ISBN: 0-672-22476-3
**No. 22476, $21.95**

## Printer Troubleshooting & Repair

*John Heilborn*

The computer hobbyist with a basic understanding of electronics can learn to service, troubleshoot, and repair a printer with this easy-to-use guide. It includes all types of printers—dot matrix, ink jet, and laser.

Preventive maintenance is emphasized with common malfunctions and repair procedures highlighted. Well-illustrated coverage is complete with troubleshooting diagrams and detailed drawings.

Topics covered include:

- A Look at Printer Types: How They Work
- Troubleshooting Techniques
- Print Heads
- Ribbon, Paper, and Print Head Transport Mechanisms
- Power Systems
- Logic and Control Systems
- Laser Printers
- Preventive Maintenance
- Appendices: IC Pin Diagrams and List of Suppliers

John Heilborn is a computer book author in the San Francisco Bay Area with approximately 20 books to his credit.

200 Pages, 8½ x 11, Softbound
ISBN: 0-672-22543-3
**No. 22543, $19.95**

## Troubleshooting with the Oscilloscope, Fifth Edition

*Robert G. Middleton,*
*Revised by Joseph J. Carr*

This 225-page book will help technicians and hobbyists obtain the maximum benefits from the wide selection of scopes now available for servicing and repair.

The new edition emphasizes digital troubleshooting and applications as well as the time- and data-domain analyzers and the logic state analyzer, with up-to-date information about television receiver troubleshooting.

Topics covered include:

- Introduction to the Oscilloscope
- How to Operate an Oscilloscope
- Using Oscilloscope Probes
- TV Signal Tracing in RF, IF, and Video Amplifiers
- Signal Tracing in the TV Sync Section
- Troubleshooting in the TV AFC and Horizontal Oscillator Section
- Waveform Tests in the Horizontal Sweep Section
- Troubleshooting in the TV Vertical Sweep Section
- Signal Tracing in the TV Sound-IF and Audio Sections
- Special Oscilloscopes and Their Applications
- Digital Logic Troubleshooting with the Oscilloscope

225 Pages, 5½ x 8½, Softbound
ISBN: 0-672-22473-9
**No. 22473, $16.95**

## John D. Lenk's Troubleshooting & Repair of Audio Equipment

*John D. Lenk*

This manual provides the most up-to-date data available and a simplified approach to practical troubleshooting and repair of major audio devices. It will enable both the beginning and the intermediate level technician or hobbyist to apply tips and tricks to any specific equipment.

This book also includes such time-saving hints as circuit-by-circuit troubleshooting based on failure or trouble symptoms, universal step-by-step procedures, and actual procedures recommended by manufacturers' service personnel.

Topics covered include:

- Introduction to Modern Audio Equipment Troubleshooting
- Troubleshooting and Repair of Amplifiers and Loudspeakers
- Troubleshooting and Repair of Linear-Tracking Turntables
- Troubleshooting and Repair of Audio Cassette Decks
- Troubleshooting and Repair of AM/FM Stereo Tuners
- Troubleshooting and Repair of CD Players

John D. Lenk is the author of over 60 books in electronics. Over one million of his books have been sold in six different languages.

208 Pages, 8½ x 11, Softbound
ISBN: 0-672-22517-4
**No. 22517, $21.95**

**Visit your local book retailer, use the order form provided, or call 800-428-SAMS.**

|  |  |  | 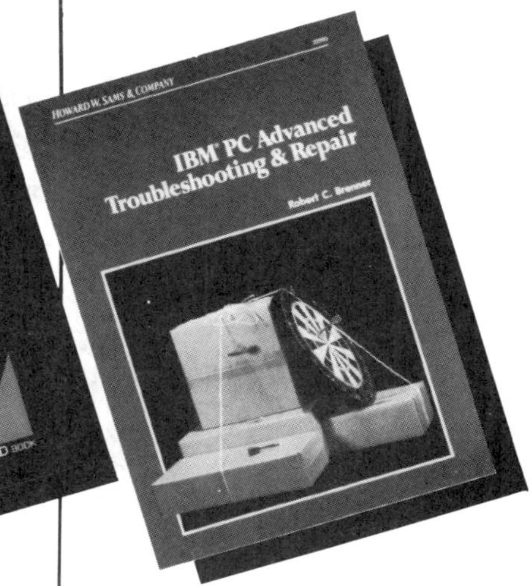 |
|---|---|---|---|

## IBM® PC Peripheral Troubleshooting and Repair

*C. J. Brooks*

Technically oriented personal computer owners, service technicians, students, and hobbyists can learn to service and repair computer peripherals with this step-by-step procedures manual.

Using a cookbook approach, this book presents the basic theory of operation, hardware descriptions, preventive maintenance, and troubleshooting and repair procedures for IBM computer peripherals. Flowcharts, block diagrams, lists of replacement parts, and many illustrations make basic repairs clear.

Reference is made to COMPUTERFACTS™ technical service data from Howard W. Sams when relevant.

Topics covered include:

- Basic Computer Organization
- Troubleshooting Basics
- Disk Drives/Hard Disk Drives/Floppy Disk Drives
- Monitors: IBM Monochrome
- Input Devices: Keyboards, Keypads, and Pointing Devices
- Printers/Impact Printing/Non-Impact Printing
- Modems

320 Pages, 8½ x 11, Softbound
ISBN: 0-672-22529-8
**No. 22529, $19.95**

## Apple® II Plus/IIe Troubleshooting & Repair Guide

*Robert C. Brenner*

This manual is structured to make it quick and easy for either the novice or the experienced technician to locate and correct most computer failures.

It introduces troubleshooting and repair in easy-to-understand terms and proceeds step-by-step through the diagnostic techniques for microcomputer hardware.

The appendices provide a wealth of backup information including step-by-step disassembly and reassembly instructions, conversion tables, specification data sheets, and system configuration record and repair sheets.

Topics covered include:

- Basic Troubleshooting
- The Apple II Plus Described
- Apple II Plus Operations
- Specific Troubleshooting and Repair, Apple II Plus
- Apple IIe Described
- Apple IIe Operations
- Specific Troubleshooting and Repair, Apple IIe
- Routine Preventive Maintenance
- Advanced Troubleshooting

256 Pages, 8½ x 11, Softbound
ISBN: 0-672-22353-8
**No. 22353, $19.95**

## IBM® PC Troubleshooting & Repair Guide

*Robert C. Brenner*

Keep your IBM PC in top operating condition with this handy reference. Its pages of schematics, photos, and block diagrams will help to identify problems and zero-in on the malfunctioning component quickly, for reduced down-time.

Easy-to-understand circuit diagrams and simple instructions written in a clear, concise style allow you to make most repairs with few or no tools.

Technicians, hobbyists and even computer novices will feel comfortable with the complex world of electronic troubleshooting. Discover how learning can be interesting and rewarding.

Topics covered include:

- The IBM PC Described
- IBM PC Operations
- Basic Troubleshooting
- Specific Troubleshooting & Repair for the IBM PC
- Preventive Maintenance
- Advanced Troubleshooting Techniques
- Appendices: IBM PC Data Sheet, Chip Information Chart, System Unit Disassembly Instructions, System Unit Reassembly Instructions, Keyboard and Disk Drive Disassembly and Reassembly Instructions, Replacing Surface-Mounted Components on the System Board, Summary of Cautions and Notes, ASCII Code Chart, and a Hexadecimal-to-Decimal Conversion Chart

208 Pages, 8½ x 11, Softbound
ISBN: 0-672-22358-9
**No. 22358, $19.95**

## IBM® PC Advanced Troubleshooting & Repair

*Robert C. Brenner*

Here is in-depth, advanced data for troubleshooting and repair of the IBM PC. This technical book compliments the circuit-tested COMPUTERFACTS™ service data with its descriptive text and expanded circuit explanations.

This book provides more sophisticated data than the author's earlier text, *IBM® PC Troubleshooting & Repair Guide.* It details PC operation and gives a full description of how to use test equipment with PC repair. Troubleshooting programs help locate failures while numerous oscilloscope screen photos and drawings help to identify problems.

Topics covered include:

- System Overview
- Detailed System Operation
- Troubleshooting Techniques
- Preliminary Service Checks
- Detailed Circuit Troubleshooting/-Analysis: Power Supply, System Board, Monochrome Monitor/Printer Adapter, Color Graphic Monitor Adapter, Keyboard, Disk Drive Adapter, and Disk Drives (A & B) Type 1
- Appendices: Disassembly Instructions, Reassembly Instructions, Adjustments to System Board and Power Board, Schematic Notes, Switches & Jumpers, and Safety Precautions & Warnings.

304 Pages, 8½ x 11, Softbound
ISBN: 0-672-22590-5
**No. 22590, $24.95**

**Visit your local book retailer, use the order form provided, or call 800-428-SAMS.**

### Commodore 128®
### Troubleshooting & Repair

*John Heilborn*

This book is a step-by-step "how-to" manual written for the owner or operator of a Commodore 128. From hobbyist to service technician, users at all levels will find the information necessary for properly maintaining and repairing the C-128 computer.

Another in the successful consumer-oriented troubleshooting and repair series for Sams, this book covers diagnostic techniques, operation, hardware, preventive maintenance, required tools, data sheets, assembly and disassembly instructions, and a comprehensive glossary.

Topics covered include:

- A Look at the Commodore 128
- Commodore 128 Systems and Subsystems
- Troubleshooting Techniques
- Digital Logic Circuits
- Troubleshooting a System That Doesn't Work
- Troubleshooting a Poor Video or No Video
- Troubleshooting a Defective Keyboard
- Troubleshooting Input/Output Problems
- Preventive Maintenance
- Appendices: Integrated Circuit/Pin-outs, Board Layout, I/O Ports, System Specifications, and Commodore Keyboard Codes

160 Pages, 8½ x 11, Softbound
ISBN: 0-672-22554-9
**No. 22554, $19.95**

### VCR Troubleshooting & Repair Guide

*Robert C. Brenner*

With approximately 25 million video cassette recorders on the market, this long-awaited book will help owners repair these popular machines when they break down.

This helpful troubleshooting guide is for the electronics hobbyist, layperson, or technician who needs a preventive maintenance and troubleshooting reference for VCRs. Limited electronics experience is required to use it, but more sophisticated service and repair functions are discussed, and valuable information for the service technician is included.

Topics covered include:

- Introduction
- Principles of Video Recording
- The VCR System Described
- Basic VCR Troubleshooting Techniques
- Troubleshooting Specific Problems
- VCR Preventive Maintenance
- Appendix and Glossary of VCR Terms

Robert Brenner is a specialist in microelectronic applications as well as troubleshooting and repair of electronic equipment. He is also the author of the best-selling troubleshooting and repair books for the Apple,® IBM,® and Commodore 64.®

256 Pages, 8½ x 11, Softbound
ISBN: 0-672-22507-7
**No. 22507, $19.95**

### Microwave Oven Troubleshooting & Repair

*John Heilborn*

Written for microwave oven owners and service technicians, this step-by-step guide will help you locate and correct problems to restore your oven to proper working order.

This unique guide provides quick, straight answers to the common problems of microwave ovens and includes all major brands. Malfunctions are discussed in easy-to-understand terms and repair procedures are explained. The book includes wiring diagrams of major brands and numerous illustrations and charts for troubleshooting.

Topics covered include:

- Microwave Oven Features
- Troubleshooting Techniques
- Components and Subsystems: Making Repairs
- Troubleshooting Guide
- Preventive Maintenance
- Product Data Sheets
- Schematics
- Component Subassembly Guides
- Replacement Parts Guide

160 Pages, 8½ x 11, Softbound
ISBN: 0-672-22481-X
**No. 22481, $16.95**

### Understanding Digital Troubleshooting, Second Edition

*Don L. Cannon*

Digital electronic systems are more reliable than the systems they replace, yet, at some point, they will need repair and maintenance. This book provides an insight into this highly technical world in a language that both technicians and non-engineers can understand. It presents the basic concepts and fundamental techniques needed to locate faults in digital systems and how to repair them.

Topics covered include:

- Digital Systems Fundamentals
- Digital Systems Functions
- Troubleshooting Fundamentals
- Combinational Logic Problems
- Sequential Logic Problems
- Memory Problems
- Input/Output Problems
- Basic Timing Problems
- Advanced Techniques

272 Pages, 7 x 9, Softbound
ISBN: 0-672-27015-3
**No. 27015, $17.95**

**Visit your local book retailer, use the order form provided, or call 800-428-SAMS.**